Frank Joseph Wethered

Medical microscopy

A Guide to the use of the Microscope in medical Practice

Frank Joseph Wethered

Medical microscopy
A Guide to the use of the Microscope in medical Practice

ISBN/EAN: 9783337076702

Printed in Europe, USA, Canada, Australia, Japan

Cover: Foto ©berggeist007 / pixelio.de

More available books at **www.hansebooks.com**

LEWIS'S PRACTICAL SERIES.

LONDON: H. K. LEWIS, 136 GOWER STREET, W.C.

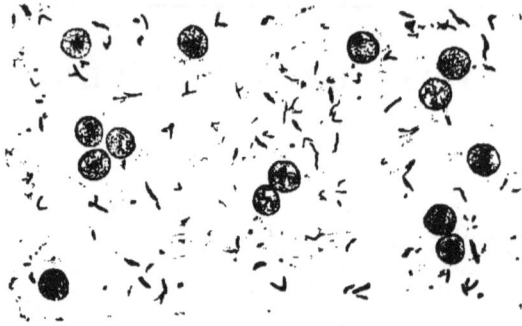

Fig. 1. TUBERCLE BACILLI IN THE SPUTUM
Stained by Neelsen-Ziehl method (from Jaksch).

Fig. 2. LEPTOTHRIX BUCCALIS
Stained with iodine solution (from Jaksch).

MEDICAL MICROSCOPY

A GUIDE TO THE USE OF THE MICROSCOPE
IN MEDICAL PRACTICE

BY

FRANK J. WETHERED, M.D. (Lond.)

MEMBER OF THE ROYAL COLLEGE OF PHYSICIANS; MEDICAL REGISTRAR TO THE
MIDDLESEX HOSPITAL AND DEMONSTRATOR OF PRACTICAL MEDICINE
TO THE MIDDLESEX HOSPITAL MEDICAL SCHOOL; LATE
ASSISTANT PHYSICIAN TO THE CITY OF LONDON
HOSPITAL FOR DISEASES OF THE CHEST,
VICTORIA PARK

WITH ILLUSTRATIONS

LONDON
H. K. LEWIS, 136 GOWER STREET, W.C.
1892

PREFACE.

WITHIN recent years the microscope has gradually been brought more and more into use as an aid to diagnosis. In certain classes of diseases, notably those of the chest and urinary organs, it has become almost a routine practice, in hospitals at any rate, to examine the sputa and deposits, and the use of the microscope is rapidly being extended to private practice.

In this work an effort has been made to lay before practitioners and students the most simple methods of preparing microscopic sections, and of the examination of urinary deposits, sputa, blood, &c. At the same time, for those who wish to extend their studies further, more elaborate methods have been detailed, and chapters added on the examination of food and bacteriological methods. It would have been beyond the scope of the book to have given descriptions of all the pathological changes found in the different organs, but a chapter on "Tumours" has been inserted, as for purposes of diagnosis, such examinations are frequently made in the operating theatre or post-mortem room.

Many of the illustrations are original, but I have also to thank Profs. Ziegler, von Jaksch and Crookshank,

Drs. Troup, Parkes and Crocker, for permission to use many of their figures.

I have also to thank Prof. Stirling, Drs. Vincent Harris and Sims Woodhead, for allowing me to copy many formulæ from their works, and especially Dr. Wynter for consenting to my making use of the material collected for our joint work on "Clinical and Practical Pathology." Also my thanks are due to Mr. Mansell Moullin, Mr. Eve, Dr. A. Garrod, and Dr. Herman, for their kind advice and help, also to Mr. Lewis for his aid in the matter of illustrations, and to the various firms who have provided *clichés* of apparatus.

Queen Anne Street,
July, 1892.

CONTENTS.

CHAPTER I.

THE MICROSCOPE AND ACCESSORIES.

CHAPTER II.

HARDENING AND DECALCIFYING.

CHAPTER III.

EMBEDDING.

CHAPTER IV.

CUTTING SECTIONS.

CHAPTER V.

STAINING.

CHAPTER VI.

SELECTIVE STAINS.

CHAPTER VII.

CLEARING AND MOUNTING.

CHAPTER VIII.

COMPLETE PROCESSES FOR THE PREPARATION OF SECTIONS.

CHAPTER IX.

METHODS OF PREPARING SECTIONS OF THE CENTRAL NERVOUS SYSTEM.

CHAPTER X.

INJECTION OF TISSUES.

CHAPTER XI.

EXAMINATION OF FRESH TISSUES.

CHAPTER XII.

PREPARATION OF INDIVIDUAL TISSUES AND ORGANS.

CHAPTER XIII.

THE EXAMINATION OF TUMOURS.

CHAPTER XIV.

EXAMINATION OF URINARY DEPOSITS.

CHAPTER XV.

THE EXAMINATION OF THE FÆCES.

CHAPTER XVI.

Examination of Sputum.

CHAPTER XVII.

The Micro-organisms of Sputum.

CHAPTER XVIII.

Examination of Vomit.

CHAPTER XIX.

Examination of Discharges and Contents of Cavities.

CHAPTER XX.

Discharges from the Genital Organs.

CHAPTER XXI.

Examination of the Blood.

CONTENTS.

CHAPTER XXII.

CUTANEOUS PARASITES.

PAGE

Vegetable parasites: Trichophyton tonsurans, Microsporon furfur, Achorion Schönleinii — Animal parasites: pediculi, Sarcoptes hominis, Acarus folliculorum 350

CHAPTER XXIII.

EXAMINATION OF FOOD AND WATER.

Wheat—Barley—Rye—Oatmeal—Maize—Rice—Beans—Peas— Sago—Tapioca—Arrowroot—Potato—Turmeric—Bread— Flour—Tea—Coffee—Cocoa—Examination of drinking water 358

CHAPTER XXIV.

BACTERIOLOGICAL METHODS.

General considerations—Apparatus required: steam-steriliser, hot-air steriliser, hot-water filter, incubator, accessories— Cultivating media—Solid media: potato cultivations, nutrient gelatine, nutrient agar-agar, glycerine agar-agar, sterilised blood serum—Liquid media: bouillon, liquid blood serum, Pasteur's fluid, Cohn's fluid—Mode of using the solid nutrient media: test-tube cultivations, plate cultivations—Mode of using the liquid media: drop cultures— Examination of micro-organisms: cover-glass preparations, cover-glass impressions—Experiments on animals . . 373

LIST OF ILLUSTRATIONS.

MEDICAL MICROSCOPY.

CHAPTER I.

THE MICROSCOPE AND ACCESSORIES.

THE first question that naturally arises when preparing to study the use of the microscope in medicine is as to the **choice of an instrument.** The number of different forms of microscopes now in the market is legion, and it is an exceedingly difficult matter to make a selection of any particular one. In order to do so, an intimate knowledge of its structure and mode of working is necessary, and it is very advisable that an inexperienced person, instead of choosing for himself, should seek the advice of a more expert friend to help him in his search, for nothing is more annoying, after expending a large sum on a large showy stand, than to find that for practical scientific work it is almost useless.

Before stating in detail the requisites of a good microscope, it will be well to give a short description of the instrument as sold by all modern makers.

Microscopes are distinguished as **simple** or **compound.** In the former, the rays which enter the eye of the observer come from an object brought near to it after refraction through either a simple lens, or a combination of lenses, acting as a simple lens, its action as a " magnifier " depending on its enabling the eye to

B

diaphragms, and by means of screws can be moved laterally, or upwards and downwards.

A second "condenser" is usually supplied for purposes of looking at opaque objects by reflected light; this, however, is seldom, if ever, required for medical work, all the objects being transparent and examined by transmitted light obtained from below.

The column is connected with the base by means of a stiff hinge, so that the tube can be fixed either in a perpendicular or horizontal position, or at any angle.

What then are the requisites of a good microscope?

For useful practical work, it is desirable not to have too large an instrument. Especially should a binocular be avoided, it is cumbersome, the number of screws around the stage becomes bewildering, and no advantage is to be gained by it for medical work over a well made mon-ocular, with a simple sub-stage condenser.

Steadiness is essential, this is secured by a wide base; in nearly all microscopes the pedestal is either of the horse-shoe, or some modification of the tripod pattern; it should be heavy enough to allow of both adjustments being turned, and of the object being freely moved on the stage without shaking the instrument.

The **column** must of course be perfectly rigid, and this property is secured by the use of good stout metal.

Both **adjustments** should turn easily and regularly, especially the fine one, for if this be at all stiff, or if the thread be in the least irregular, great annoyance will be experienced.

The **stage** is usually three inches long by two and a half wide. It is not advisable to have a "movable" stage. By means of the fingers the glass slide can be moved more readily, and by practice quite as smoothly as by any mechanical contrivance, whilst extra screws and levers are apt to be in the way.

There should be plenty of room between the mirror and the stage, so as to allow the insertion of a condenser, if sufficient space be not allowed the full effect

FIG. 2.—Baker's new student's microscope.

of this contrivance is not experienced. It is most certainly desirable to have a condenser; the search for micro-organisms has now become almost a matter of routine in certain cases of doubtful diagnosis, and for their sure detection a special illumination is absolutely necessary.

The **diaphragm** is made of several different patterns. The most ordinary form consists of a "wheel" of diaphragms of various sizes, and should be examined to see whether the apertures are concentric with the axis of the instrument. Another form consists of small blocks in which the openings are pierced, all fitting into a common receptacle, which is inserted below the stage. The most important feature is that the diaphragm must be placed *immediately* beneath the object, in order to secure the best results.

There are several more complicated forms such as the "iris," but for a description of them the reader is referred to larger treatises on microscopy.

Although the choice of a "stand" is of no mean importance, the chief point is of course the **selection of the objectives.**

For ordinary histological work, an 'inch' lens and a 'quarter-inch' is sufficient, but if it be intended to undertake the examination of sputum, &c., a twelfth-oil-immersion is necessary, as the micro-organisms cannot be seen with any certainty without this power.

According to the Continental systems, the first named lenses are represented by Nos. 3 and 7 of Hartnach, or A and D of Zeiss.

It should be borne in mind that the term "one-inch objective," &c., indicates only that such a lens should possess the same magnifying power as a single lens of one-inch focus, and that it does not refer to the distance which the front of the lens focusses from the object, this varying greatly with the different makers.

The **qualities of a good objective** are:—defining power, flatness of field, penetrating power, and resolving power.

Defining power.—Expressed in more exact lan-

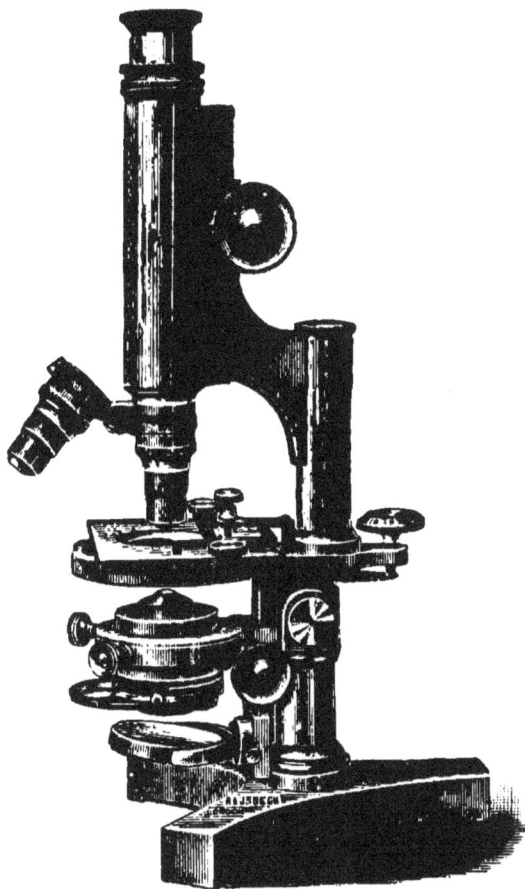

FIG. 3.—Beck's pathological microscope.

guage this means that the objective has been properly corrected for spherical and chromatic aberration.

"**Spherical aberration**" causes the lines of the object to appear blurred and foggy, instead of being distinctly defined and clearly separated from one another. If the glass be not free from **chromatic aberration,** coloured fringes appear around certain spots, blue, if the lens be under-corrected, red, if over-corrected. "Defining power" may be tested by examining a stage-micrometer and observing if each division can be clearly distinguished.

Flatness of field.—If an object be moved about from one portion of the field to another, it ought to appear equally well-defined in all parts. Thus, taking the stage-micrometer again as a test, the lines should be sharp and clear, and perfectly parallel to one another in every part of the field.

Penetrating power.—By this term is signified the power of an objective to view several planes of an object without altering the focus. A section of liver is a good test for judging this properly.

The penetrating power of a lens is dependent on its "**angle of aperture.**" This is the angle made by two lines drawn from opposite sides of the aperture of the object-glass with the principle focus of lens.

If the system of lenses has a "low angle," its penetrating power is greater than one which possesses a "high angle" of aperture.

In connection with this subject it may be mentioned that low-power objectives with a high-angle and consequent shorter working distance will define much better than those with smaller apertures.

Resolving power.—This again depends entirely upon the aperture, supposing the glass to be perfect in other respects; for good effects the angle ought to be large. Some authorities object to the term "resolving

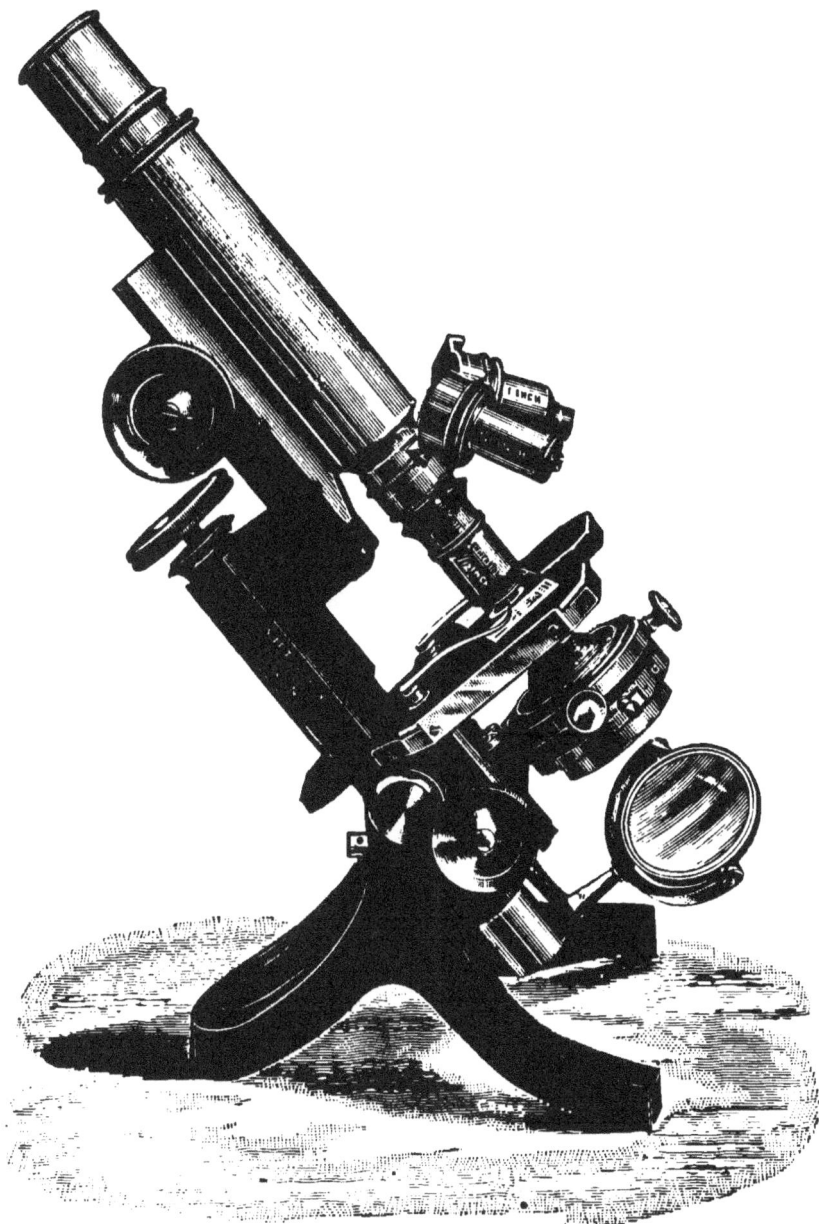

Fig. 4.—Swift's bacteriological microscope.

power " as being synonymous with "defining power."
An objective, however, may have perfect defining power ʎ
and yet by reason of its want of aperture, it will be unable to distinguish certain minutiæ.   It defines all that
it can take up, but cannot define what is not imaged
by it.

In choosing an objective, it is well to select one with
an **universal screw,** so that it can be used with any
stand, otherwise an "adapter" will become necessary.

A **"nose-piece"** (fig. 5) carrying two or three objectives is a useful adjunct, saving a great deal of time

FIG. 5.—Nose-piece.

when working with more than one power.   This little
piece of apparatus should be tested to ensure of there
being sufficient space for the oil-immersion to revolve
between the tube and the stage, and some minute object
should be selected by means of the low power and
arranged so that it occupies the centre of the field ; the
high power should then be turned round, and lowered
on to the glass, when this particular object should still

be in the centre of the field; if this be not the case, the nose-piece is not correctly "centred" and should be rejected.

Not more than two **oculars,** a high and low, will be required; their function, as already stated, is to magnify the image formed by the objective; but enlargement is better obtained by drawing out the tube, or by substituting a stronger lens, than by using a high eye-piece, which is trying to the eyes.

A **Camera-lucida** is also of service for drawing the outline of specimens; one need not be bought, however, for it is easy to manufacture one by bending a small piece of tin-foil round the upper end of the tube, and arranging it in such a fashion as to support a cover-glass at a proper angle to the eye-piece; the angle is readily ascertained by experiment, and this simple contrivance has the advantage over the more complicated ones, in that the glass being so thin, only one image of the pencil instead of two is seen, and it is consequently much easier to use. This is the principal of Beale's "neutral tint reflector."

It would be invidious and beyond the function of this book to recommend any particular maker from whom a microscope should be bought. There are many excellent dealers both in London and the Provinces. It is presumed that those who make use of these directions will not be desirous of possessing a large and expensive instrument. Very good "student's microscopes" are now made by many of the London firms, which can be highly recommended. They are fitted with all the apparatus necessary for simple bacteriological work, and the cost is from £16 to £20.

Any of the following makers may be fully trusted to supply good instruments, and any omissions that occur do not imply that therefore other firms are not equally as reliable :—

C. Baker, 243 and 244 High Holborn.
Beck, 68 Cornhill.
Crouch, 66 Barbican.
Pillischer, 88 New Bond Street.
Powell and Lealand, 170 Euston Road.
Swift, 81 Tottenham Court Road.
Parkes, 5 St. Mary's Row, Birmingham.
Ross, 164 New Bond Street.

In America very similar instruments, possessing everything that can be desired, and excellent both in manufacture and in the qualities of the lenses, may be obtained from Joseph Zentmayer, 209 South Eleventh Street, Philadelphia, and Bausch and Lamb, Optical Company, 48 and 50 Maiden Lane, New York, and at 543 North St. Paul Street, Rochester, N. Y. These firms also supply all microscopical apparatus and accessories.

If a suitable stand be already to hand, the extra apparatus may of course be added. This applies more particularly to an oil-immersion lens and Abbé's condenser. The best immersions are perhaps those of Zeiss, but they are very expensive. A very good lens is made by Leitz, and costs about £5, whilst a condenser can be fitted for about £3. Both the lens and condenser may be obtained from any of the makers above mentioned.

Having now detailed the form of the microscope and the essentials of a good and practical one, we now pass to the **use of the instrument.** This may seem somewhat superfluous, but a few hints to the inexperienced may save them not only various petty annoyances, but probably also the loss of valuable lenses.

In removing the instrument from its box it should be grasped by the column, not by its foot, as thereby much

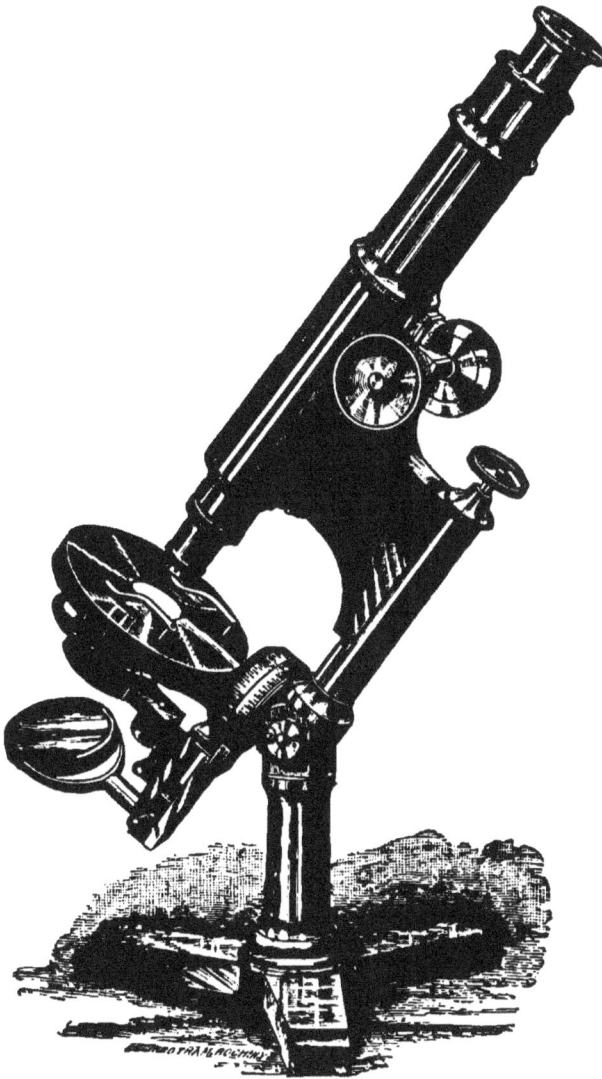

FIG. 6.—Bausch and Lamb's universal microscope.

strain on its bearings is prevented. Any jarring of the adjustments soon puts them out of order. For this reason when in constant use it is preferable to keep it under a bell-jar rather than to be constantly lifting it out of and into its case. Scrupulous cleanliness cannot be too strongly insisted upon. Each time the lenses are used they should be carefully wiped with a soft piece of chamois leather. This applies more particularly to the oil-immersion lens, for if the oil be allowed to remain on any length of time it is apt to become hard, and great trouble is experienced in removing it.

It not infrequently happens when working with the high powers that these are accidentally forced through the cover-glasses and so become covered with balsam; this is best removed by moistening the chamois leather with a little alcohol or xylol, rapidly passing it over the lens and immediately drying it. This must be done very cautiously as the lenses are "set" in balsam, and there is some danger of this being dissolved by the spirit, and if such be the case, the damage is extremely difficult to rectify.

Having settled the microscope on a firm table or stand (for any vibration is very disturbing) and inclined it at an angle to suit the convenience of the observer, without causing him to adopt an uncomfortable or tiring posture, the mirror is turned so as to reflect the light up the tube. The best illumination is obtained with a North light, so as to avoid the direct glare of the sun which is trying to the eyes; a very good light is obtained by reflection from a white cloud. If a low power is being used the flat side of the mirror should be employed, and if a high one, the concave side.

In conjunction with the mirrors, various diaphragms

are used to regulate the quantity of light passing to the specimen. In examining unstained sections, or objects such as are found in urinary deposits, the diaphragm with the smallest aperture that can illuminate the whole of the ·field is applied. With stained specimens no regulation of the amount of light is necessary.

With regard to the selection of a power to view any particular specimen, the rule is, never use a higher power than is absolutely necessary for clearly bringing out the minutiæ of the object under examination. The inch lens should therefore first be used to gain a general idea of its structure and nature, and then the quarter-inch may be substituted; for ordinary medical work a higher power will not often be needed, except for the examination of micro-organisms. In using the high powers, in order to avoid breaking the cover-glass by injudicious use of the coarse adjustment, it is a good plan before applying the eye to the tube to lower the lens until it just touches the cover-glass, and then to work it upwards by means of the fine adjustment whilst looking down the microscope; if the structure be very fine, it is advisable to slightly move the slide up and down with the left hand, as a moving object is more easily caught sight of than one which is still.

It should be remembered that the magnifying power may be augmented:—(1) by increasing the power of the object-glass; (2) by using a higher eye-piece; (3) by increasing the distance between the object-glass and the eye-piece, that is by drawing out the tube.

In using the oil-immersion a drop of thick cedar oil is placed on the cover-glass, and the lens lowered until it comes into contact with the oil. The fine adjustment is then cautiously turned until the object comes into focus. At the same time care must be taken that the

oil does not flow over the cover-glass and dissolve the
balsam in which the section is mounted ; to prevent this
it is a good plan to ring the specimen with gold size.

The Abbé's condenser (fig. 7) must be focussed so as
to produce the brightest possible illumination.   The
principle of the condenser is this :—It is of such a form
that the rays of light reflected from the mirror pass
through it in such a direction that their convergence
forms a cone of light as highly concentrated as possible.

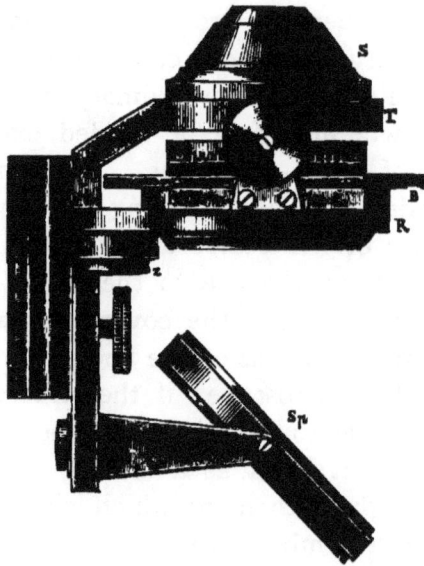

FIG. 7.—Abbé chromatic condenser.

Under these circumstances the outlines of unstained preparations appear blurred, and this illumination is consequently unsuited for fresh specimens.   Also, with stained
preparations the structure of the tissue or whatever
may be under examination cannot be distinctly made
out, but the portions which are more deeply stained,
such as the nuclei, are brought clearly into view.   It is

for this reason that the condenser is so valuable in searching for micro-organisms; after the use of suitable stains the small rods and micrococci stand prominently out from the indistinctly defined back-ground and thus can readily be seen and recognised.

If diaphragms are placed between the mirror and condenser the characteristic action is cut off, and a good illumination is obtained for histological work and also for the examination of unstained micro-organisms, as in "drop cultures." The same effect is also produced by lowering the condenser by means of screws provided for the purpose.

A caution may be given here on the use of the nose-piece. In causing this mechanism to revolve, care must be taken to raise the tube high enough to allow the high powers to travel without touching the glass, otherwise, of course, the object will be moved, and thus the great advantage of the nose-piece frustrated. This is par-ticularly liable to occur after searching the specimen for any one spot with a lower power, which it is desired to investigate with the oil-immersion lens, as in exam-ining tissues for tubercle bacilli, &c.

The two clips may be removed for ordinary work, but when using the immersion lens, it is a good plan to take out the left-hand one only, leaving the right to fix one end of the slide whilst the other is moved about with the fingers.

The chief use of the clips is to fix the slide in one position either for purposes of demonstration or when drawing with or without the camera-lucida.

It is not always possible to work by daylight, and consequently some form of artificial light must be em-ployed. Nothing equals an oil-lamp. Various kinds of "microscopic lamps" (fig. 8) are sold, which have in

c

themselves no special advantages except in conjunction with very high powers, when by means of a slit in the shade, only a thin stream of light is allowed to fall on the mirror, so producing more intense but localised illumination. For purposes of laboratory work at night, as well as for use with the microscope, " The Queen's " reading lamp can be confidently recommended. The

FIG. 8.—Microscopic lamp.

common colza or any mineral oil may be burnt; it is excellent as regards brilliancy and steadiness of light.

With the modern improvements in electric-lighting it is probable that this mode of illuminating dwelling-houses will soon become more general, and this pure white light is very good for microscopic work, although it is often rather difficult to avoid the reflection of the incandescent wires.

A good deal of discussion has taken place as regards the correct manipulation of the mirror, especially when working with the condenser as reflections of the window or of the lamp are extremely apt to appear in the field. The only way to remedy this is to turn the mirror so as just to cause these images to escape the circle of the lens, this giving the next best light to the centre of convergence of the rays.

The next point to be considered in connection with the use of the microscope is the representation on paper of what is seen through the tube. This is naturally an important matter, as it is almost impossible to give such a graphic description of the microscopical specimen as shall enable everyone who reads it to make a mental picture of what the writer means. Time is not wasted, therefore, which is spent practising drawing objects, for the pains taken will be amply repaid.

The most simple method is to place a sheet of paper on the right of the microscope, and then to look down the tube with the left eye, keeping the right open gazing at the paper; an image of the object will then appear to be thrown on the paper, and the outlines may be traced with a pencil. If the paper be placed on a level with the stage, the picture when drawn will be of the same size as the image seen through the microscope. If it be placed higher, the representation will be smaller, whilst if lower, then the drawing will give an enlarged outline of the field.

A more exact copy is obtained by aid of the camera-lucida. There are several different forms of this instrument; the cheapest and most simple has already been described on p. 11, and this contrivance will answer all the purposes of the more complex ones as sold by the different makers. In using it, the glass slide is fixed in position by means of the clips, and the object accurately focussed. The tube is then inclined at right angles to the stand, and the mirror arranged so that the object is illuminated, but not too brightly. The paper, placed so that the projected image shall

FIG. 9.—Prismatic camera-lucida.

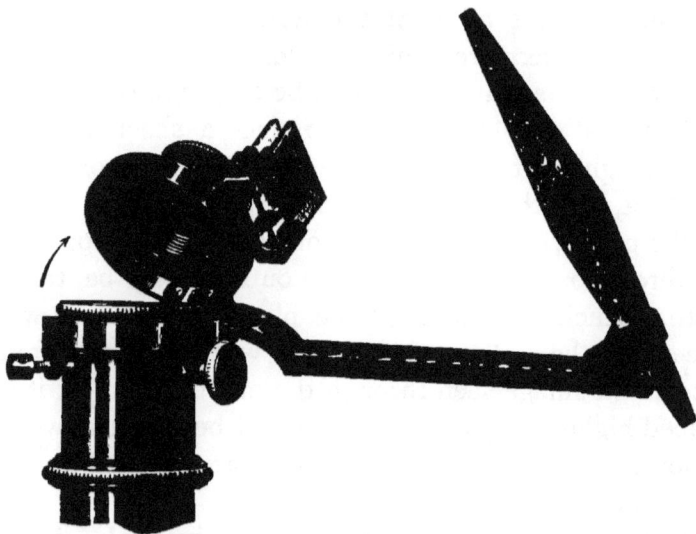

FIG. 10.—Abbé's camera-lucida.

fall upon it, should not be too much in the shade. Only the principle outlines need be traced, the shading and minutiæ being filled in freehand. The drawing is best made with a hard pencil.

Two other forms of the camera-lucida are shown in the accompanying figures. Fig. 9 is a simple prismatic arrangement, but quite effectual. Fig. 10 is the instrument introduced by Abbé, and possesses the advantage that it may be used with the microscope in an upright condition.

Another operation which is, however, but rarely required in ordinary medical work, is ascertaining the size of an object. An appliance is employed known as a "micrometer." The form in most common use is the "stage-micrometer" which consists of a glass slip on which is ruled a series of lines $\frac{1}{100}$ of an inch apart, one of these divisions being again divided into ten parts.

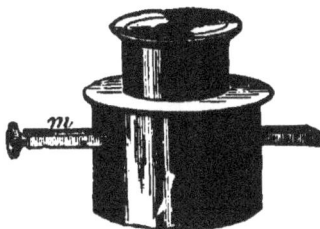

FIG. 11.—Eye-piece micrometer.

The most simple method of measuring an object is by means of a cover-glass made to act as a camera-lucida as above described. A stage-micrometer is placed under the microscope and focussed; the instrument being horizontal, a piece of paper is laid on a level with the stage, and the lines magnified by a quarter-inch lens are carefully traced. The micrometer is now removed and replaced by the object whose size it is desired to ascertain. The object is traced, and compared with the scale by the aid of compasses.

Some microscopes are provided with an "eye-piece micrometer" (fig. 11), a flat piece of glass having a scale ruled upon it inserted in an ordinary ocular. The value of the divisions of the scale must first be determined for each objective by observation of a stage micrometer, and this being once done, a note should be made for future use.   In subsequently using it for measurement all that is necessary is to see how many divisions of the scale the specimen under examination covers.

A table is usually provided with every microscope designating the magnifying power of each objective with the different eye-pieces. Should this have been mislaid, or a new lens been added, the magnifying power may be determined as follows :—

The stage-micrometer is used, and its image cast by means of the camera-lucida on a sheet of paper, placed exactly ten inches from the eye.   Two lines, which are in reality $\frac{1}{1000}$ of an inch from each other, are then traced, and the distance between them on the paper accurately measured, then the magnifying power is ascertained by calculating how many times the $\frac{1}{1000}$ of an inch will go into the distance on the paper.   Thus, supposing with an inch lens the lines of the image are $\frac{2}{10}$ of an inch apart, then, $\frac{2}{10} \div \frac{1}{1000} = 200$; therefore the system (objective and eye-piece) magnifies two hundred times.

This completes a description of the use of the microscope and we now pass to consider the **accessory apparatus** which is required.

It need not be supposed that a separate room is needful for microscopic work.   A laboratory is of course to be preferred, but all that is really requisite is a firm table and a good light.   The table is best when made of deal; it should stand on four steady legs, and be of such a height that the operator may sit comfortably before it while at work.   It is essential that the edges of the table should be square, for if they are bevelled it will be found impossible to affix the microtome to the bench by means of a clamp.

For convenience in manipulating specimens it will be found a good plan to paint on the table close by the worker, two oblong patches, one black, and the other white, each being about a foot long, and six inches wide ; the former to be used when dealing with unstained sections, and the latter with coloured.

Convenient drawers or shelves should be close at hand in which to place the specimens to preserve them from dust and guard them from possible injury.

If in the room chosen for this work there be no tap and sink, a large jug for water and a receptacle for waste must be provided. Vessels made of metal should be avoided as they rapidly become corroded by acid, and therefore earthenware ones should be selected.

It is almost impossible to prepare sections without the use of some special instrument. After considerable practice, sections indeed may be cut with a razor alone, but they are not so satisfactory, and even the most skilful worker can only succeed in procuring small ones, whilst by the aid of a **microtome** large and regular specimens may be obtained. Of these microtomes there are several kinds varying in price from a few shillings to several pounds.

For ordinary use either Frazer's modification of Cathcart's microtome, or the instrument introduced by Swift may be safely recommended.

Cathcart's microtome (fig. 12) consists of a brass plate supported on a micrometer screw. The plate is set between two parallel wooden uprights on which are fixed thick glass slides. These supports are about 6 inches long and three inches high. By means of projections the apparatus may be firmly fastened to the table by clamps provided for the purpose. The cutting instrument is a large plane-like knife set in a wooden handle. It looks very clumsy but answers every purpose admirably.

FIG. 12.—Frazer's modification of Cathcart's microtome.

When it is used for preparing frozen specimens a spray apparatus is fixed beneath the clamp, the free tube being connected with a bottle of ether which is hung on the side of the microtome.

When required for " embedded " specimens the brass disk is removed, and in its place a clamp (fig. 13) is substituted for the reception of the cork, on which the substance to be cut is fixed.

Swift's microtome (fig. 14) has a large metal clamp for affixing it to the table surmounted by a thick round plate of glass in the centre of which is a small brass disk roughened on its upper surface. The spray apparatus which is also connected with a bottle for containing ether is fixed below this, and Swift has lately devised an arrangement by which a clamp may be substituted when embedded sections are being dealt with.

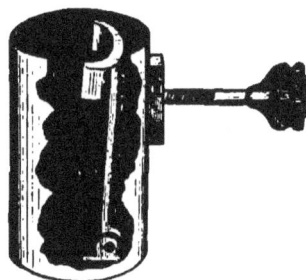

FIG. 13.—Clamp of Cathcart's microtome.

With the microtome is supplied a " plough " for holding the razor. This is a tripod metal stand supported by screws one at each angle. The one at the apex acts as a micrometer screw, and when turned, raises or lowers the edge of the knife. At the sides of the triangle are two smaller screws with notches at the lower ends. In the base of the tripod is another small screw, to which is attached a grooved block. This is for grasping the back of the razor, whilst its edge fits into the notches already mentioned. By tightening these three screws the razor is firmly fixed.

A freezing microtome in common use, especially

where a number of specimens are required for demonstration purposes, is known as **Williams' freezing microtome** (fig. 15). It consists of a round wooden

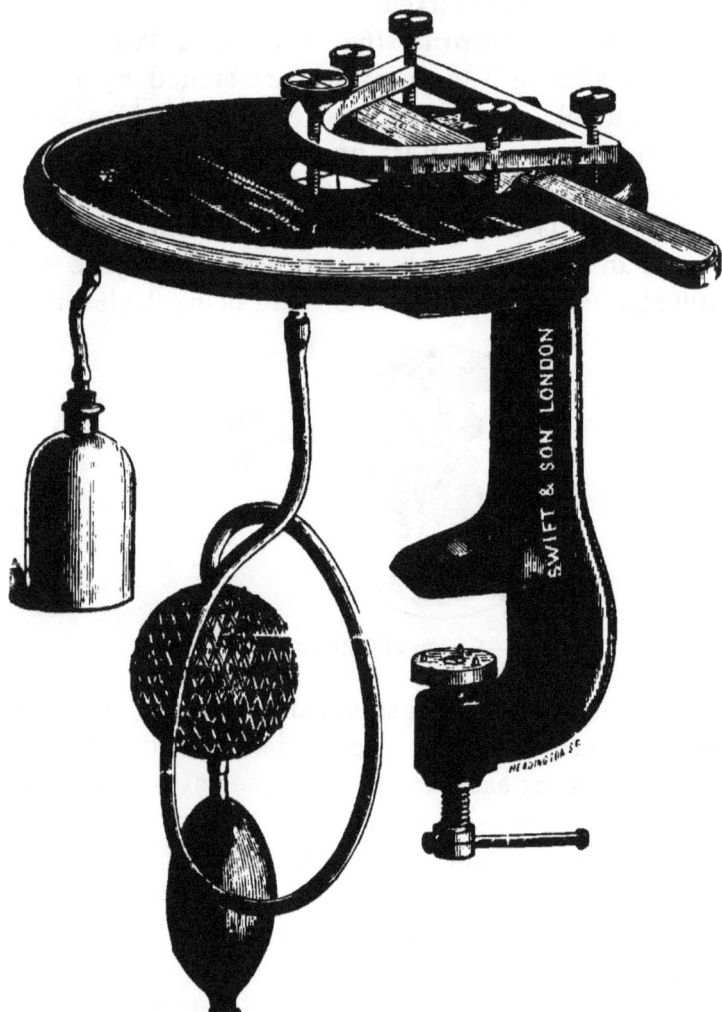

FIG. 14.—Williams' ether microtome (" Swift's microtome.")

box, provided with a drain pipe, in which the freezing mixture is placed. In the centre of the box is a vertical brass pillar, provided at the top with a roughened brass plate. The box is covered with a lid formed of a thick plate of glass, in the centre of which is a hole through which projects the plate for the reception of the specimens. The cutting instrument used in connection with this microtome is Swift's " plough " just described.

FIG. 15.—Williams' ice and salt microtome.

Other microtomes are those of Katsch, Roy, Schanze, Zeiss, Jung, &c. If it be intended to do much section-cutting one cannot do better than at once purchase the more complete instrument made by Reichert It is more expensive costing about £6, but is far better adapted for cutting " embedded " specimens.

The knife is fixed on a carriage which runs in a groove and its movement is therefore very steady. The micrometer screw is worked automatically. Its general form will be sufficiently appreciated by referring to the accompanying illustration (fig. 16). M is the razor fixed to the block K, by means of the clamp F. By knocking against the screw c, fixed by the milled head b to the carriage by means of a spring arrangement, the wheel Z is turned round and so raises the clamp g in which the specimen to be cut, is held. This clamp is fixed by the screw e. An indicator i marks the amount which the object is raised each time the lever h is set in action. a is a catch regulating the movement of the wheel Z, while W is a metal trap for catching the waste alcohol.

A very capital instrument has been introduced by the Cambridge Scientific Instrument Co., known as the **rocking microtome** (fig. 17). Its chief use is for cutting a series of sections but for ordinary purposes it has no advantage over the instruments already described.

Swift has recently introduced an ingenious modification of the rocking microtome, in which clamps are provided so that it may also be used for cutting specimens embedded in celloidin and also as a freezing microtome.

It is often useful to be able to cut sections of fresh material without any preliminary preparation, such as during an operation for the recognition of the different forms of tumour. For this purpose nothing is better than a **Valentin's knife** (fig. 18). It is formed of two narrow blades lying parallel to each other, their distance being regulated by a fine screw.

The method of using the above instruments will be described when directions are being given for cutting sections.

A certain number of accessories for convenience in

PATENT
CARL REICHERT
WIEN

W

h

microscopic work will be wanted, but these, however, require little explanation beyond enumerating those ab-

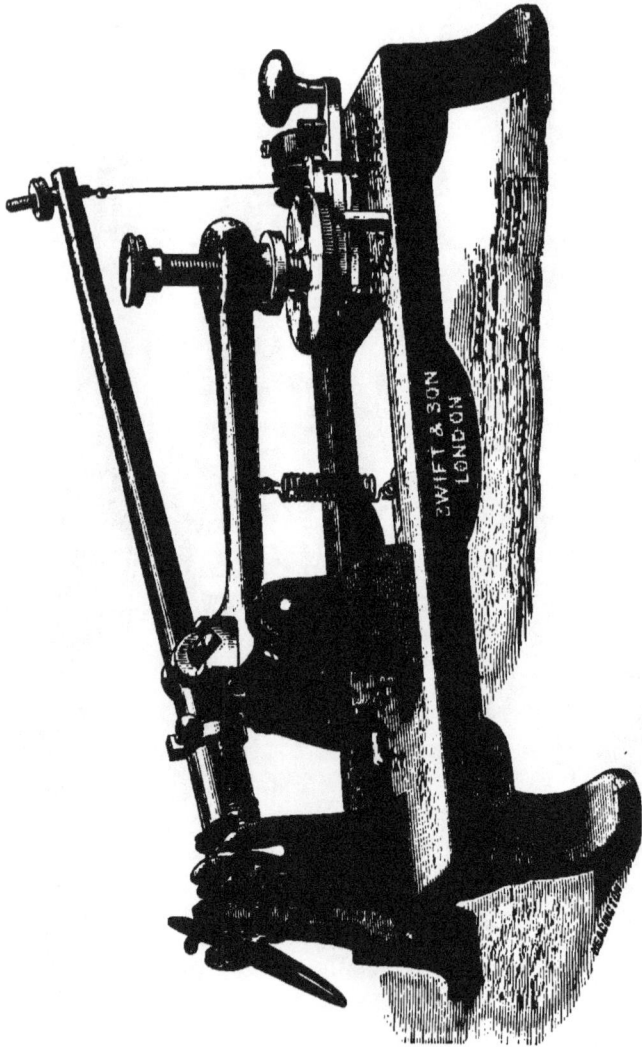

Fig. 17.—Rocking microtome.

solutely needed. Objects are nearly always examined on **glass slides** about three inches long and one inch wide.   The edges are usually ground, but in a cheaper make they are rough.   It is scarcely necessary to say that the glass should be quite free from specks and flaws of all kinds.   The specimens are covered by thin squares or circles of glass known as **cover-glasses.** These are made in varying thicknesses ; the most useful are the squares, $\frac{7}{8}$ of an inch wide, and the thickness known as " No. 2."   These glasses are sufficiently thin to be used with a $\frac{1}{4}$ inch lens and the oil immersion ; if however, an eighth dry power be used, thickness " No. 1 " must be employed.   Great care must be exercised in cleaning the cover glass, as being extremely brittle they are easily damaged.

When first purchased they are often covered with a thick film of grease which is difficult to remove.   The best mode of doing this is to place them for a short

FIG. 18.—Valentin's knife.

time in strong hydrochloric acid, then swill them in water, and finally transfer them to a covered capsule containing methylated spirit ; and here they should be kept until required.   Before use they are dried with a soft cloth, being rubbed between the finger and thumb, or a corner of the cloth is laid on the table, the cover-glass placed on it, and the glass is gently wiped on either side.   " No. 1 " cover-glasses being very thin, a good plan is to swill them in absolute alcohol as a final stage in the cleaning, this evaporates quickly, and does away with the need of a cloth.

Cover-glasses which have been used may be recleaned in the manner above described.   A tumbler or beaker covered with a ground glass plate should be kept on the bench into which all waste alcohol should be thrown. Specimens which have been mounted in balsam, but which have not been considered worth preserving can be placed in this tumbler, and cover-glasses dissolved off.

Some fine sewing **needles** should be mounted in wooden handles, and must always be kept clean and sharp. Special holders are sold in which the needles may be conveniently changed when they are blunted or spoilt.

Excellent substitutes for needles are **glass rods** drawn out to a very fine point. These are superior to needles in that they do not corrode in acids, and are not so apt to tear the specimens.

Two pairs of **forceps** will also be required ; those should be selected which have broad blades so that cover-glasses can safely be grasped by them and the student should avoid the long slender forceps which are usually sold with cases of microscopic instruments, as they are apt to chip and break the glasses.

A **scalpel** and pair of **scissors** are also necessary as also a couple of **section lifters.** The last named are preferably made of some malleable material such as stout tin foil.

Of glass apparatus the following are required. About half a dozen **capsules** of different sizes, the same number of **watch glasses** with ground bottoms, to ensure their steadiness, and two glass **funnels.**

For holding the stains when in use nothing answers better than small glass salt-cellars, they do not easily upset, and are of a convenient depth.

A packet of **filter paper** and a **filter stand** will complete the apparatus for ordinary microscopic work, whilst for the examination of urinary deposits, &c., **test-tubes,** some **conical** glasses and **pipettes** will be required.

For the reception of the reagents, bottles with glass stoppers are preferable.

A large supply of stains and other reagents is not necessary, and the following is a list of those absolutely needful. Those required for special purposes will be described as the need for them occurs, or they can always be added from time to time.

Absolute alcohol.

Methylated spirit.

Glycerine.

Farrants' solution.

Canada balsam dissolved in xylol.

Normal saline solution.

This is made in the following manner (Woodhead):—Sodium chloride is heated to redness, and cooled over sulphuric acid; 7½ parts by weight are then dissolved in 1000 parts by measure of distilled water.

Iodine dissolved in iodide of potassium.—The best formula for this is that known as "Gram's solution." This not only acts as a useful testing fluid but is also an essential part of the process for staining certain micro-organisms.

It is made as follows :—

Iodine . . . . 1 part,
Iodide of potassium . . . 2 parts,
Distilled water . . . . . 300 parts.

D

CHAPTER II.

Hardening and Decalcifying.

For examination by the microscope sections may be prepared in two ways, either directly from fresh specimens, or from those which have been duly "hardened." The former method is practically confined to the investigation of tumours in the operating theatre, or for purposes of immediate diagnosis in the post-mortem room. When prepared in this manner, however, the sections cannot be permanently mounted, and hence are only of temporary service. A knowledge of how to prepare such sections is nevertheless greatly to be desired, and a chapter will be devoted to a consideration of this subject. The processes by which specimens may be mounted, and kept for future reference, or for purposes of demonstration, will be first described.

In selecting the portions of an organ or tissue for microscopical purposes, some care has to be exercised. They should be obtained as soon after death as possible, for if decomposition has set in the sections will not properly take up the stain, and thus many important characteristics may be lost. In choosing the parts of a diseased organ, the liver for example, it is always preferable to take a particle from near the surface, so as to include the capsule, as this affords great aid in recognising the minute anatomy of the organ. Again, in preparing sections of the intestine or stomach small portions should be cut off transversely to the axis of the viscus, and about a quarter of an inch in length so as to show the condition of its various coats.

When investigating a morbid growth invading an organ at least two pieces should be cut out ; one to be taken from the centre of the tumour, and one from its circumference, so as to include a portion of the organ as well as of the growth.

The size of the pieces should be nearly an inch in length and breadth, and a little less than a quarter of an inch in thickness. They should be removed by a sharp scalpel, the cuts being cleanly made so as to avoid any tearing, and when completely freed should be lifted upon the knife, and dropped into a small wide-mouthed bottle, which should be at once labelled with the name of the organ from which the specimen was taken, the date, and supposed morbid condition. Too much stress cannot be laid upon the importance of immediately labelling all preparations whether completed or otherwise, as from postponing this simple precaution many errors are apt to arise, and cause waste of time and much annoyance. Added to which a good specimen is of little value, unless its source in addition to its nature is known.

The bottle into which the cuttings have been placed should be filled with the hardening solution, and the choice of such a solution must now be considered.

The reagents in most common use are :—

Absolute alcohol.

Methylated spirit.

Müller's fluid ; and bichromate of potash.

The selection of one of these depends upon many points.

D 2

a. The organ or tissue which is to be hardened. The most useful reagent is undoubtedly absolute alcohol. It is suitable for nearly all tissues, except those which owing to their complex structure do not contract regularly. Thus, the central nervous system, and the eye should not be subjected to its action. Thin tissues and membranes also are best hardened by some other fluid on account of the manner in which in the alcohol they curl up; for instance sections of the intestines cannot be satisfactorily obtained after hardening in this reagent. For all such exceptions Müller's fluid is preferable.

b. The nature of the morbid change. Alcohol owing to its solvent power on fat should not be used when it is desired to demonstrate fatty changes ; on the other hand it is most suitable when micro-chemical reactions are to be exhibited, such as the amyloid reaction. When specimens are hardened in Müller's fluid or bichromate of potash, such reactions are marred.

c. It is occasionally desired to investigate certain particular elements of the tissue, such as the epithelial structure, or the division of nuclei. For these purposes it is advisable to use the chrome salts.

d. For *bacteriological purposes* absolute alcohol or methylated spirit should be used, as the other reagents although they do not entirely destroy yet greatly interfere with the properties which the various bacteria possess of retaining the aniline dyes.

e. The time which the different hardening reagents occupy before the specimens are ready for cutting is occasionally of importance. If it be desired to secure a rapid and energetic action, provided there are no other objections, absolute alcohol should be chosen ; whereas if a more cautious course is advisable, Müller's fluid or bichromate of potash should be selected.

Before describing seriatim the method of using the above named hardening reagents, a few general hints may be given. A relatively large volume of the liquid should be employed and frequently changed; the periods for so doing depending upon the fluid used. As before stated the specimens should be placed in the reagent as soon as possible after removal from the body. They may either be suspended by means of threads in a tall cylindrical vessel, each thread having attached to it a label, by means of which it may be easily recognized; or a layer of cotton-wool may be placed in the jar. The reason for these precautions being, that the specimens may be entirely surrounded by the hardening reagent.

If Müller's fluid or bichromate of potash are the reagents selected, the jars or bottles should be kept in a cool dark place, as these fluids are very apt to develop moulds.

Absolute alcohol.—This acts as a hardening reagent by means of the withdrawal of water and coagulation of albumin. Specimens placed in it usually shrink somewhat, but this is of very little moment, for when the sections obtained from them are placed in water, they soon resume their natural appearance.

As previously mentioned it is suitable for nearly all structures, the exceptions being those which owing to their varied composition shrink unequally, such as the spinal cord and eye, but even these with a little manipulation may be hardened with this reagent. Thus, if it be desired to harden an eye-ball, this can be effected by injecting alcohol into the vitreous by means of a hypodermic needle before the specimen is placed in the jar, no inequalities will then be produced.

Again, lungs, muscles, and old spirit preparations are

sometimes difficult to harden. This may be accomplished by placing the specimens for twenty-four hours in equal parts of glycerine and mucilage of gum arabic, and transferring them once more to the alcohol, when an excellent consistence for cutting will be obtained.

The portions to be hardened should always be small, as owing to the rapidity with which alcohol acts, the outer surfaces of the specimens become so set as to prevent the alcohol penetrating to the interior.

The liquid generally becomes turbid a few hours after the fresh tissues are immersed in it, and therefore after twenty-four hours it is better to transfer them to a fresh lot of alcohol, but subsequently no further change need be made, as there will be no tendency to over-harden. In this respect alcohol gives far less trouble than Müller's fluid or bichromate of potash, and the hardening process is moreover more easily controlled, for the action of the last named reagents greatly depends upon time and temperature, and very often produces cloudiness and discolourations.

As regards the time required for the completion of process, this greatly depends upon the size of the specimens. Small ones will be ready in twenty-four hours, whilst larger ones will require about four days.

Methylated spirit.—This is suitable for the same class of objects as absolute alcohol, and its effects are practically the same. Its action, however, is not so rapid, and the specimens should remain immersed for about fourteen days. The spirit should be changed after the first twenty-four hours, and again after the first week.

If time be no object it is very suitable for hardening tissues in which it is intended to search for micro-organisms.

Müller's fluid.—This has the following composition :—

Bichromate of potash . . .   2 parts,
Sulphate of sodium. . . .   1 part,
Distilled water . . . . . 100 parts.

This reagent fixes the protoplasm of the cells rather
than hardens them, and so unlike alcohol causes but
little shrinking.

It may be used for any tissue except those in which
micro-organisms are to be sought for, or in which some
micro-chemical reactions may have to be made use of—
such as in staining amyloid material. There is one
other rare exception which may be mentioned, a deposit
of lime salts is dissolved by this medium, and consequently chalky deposits such as occasionally occur in
the glomeruli of the kidneys are destroyed if the organs
be hardened in this way.

Müller's fluid is especially adapted for hardening
the central nervous system, and indeed is essential
when Weigert's or Pal's methods of staining are adopted.
Portions of intestine or other hollow organs are also
best hardened by this fluid. Specimens of any size,
even whole brains, may be prepared in this manner, but
as the penetrating power of the liquid is very small, a
considerable time must elapse before the process is
complete. Even the smallest portions of any organ
require about six weeks, whilst a brain must remain
about a year. Moulds are likely to form, and to prevent this a small piece of camphor should be kept
floating in it.

The process of hardening may be hastened by placing
the jars in an oven kept at a temperature of 30 to 40° C.
(80 to 100° F.). The fluid has to be frequently changed,
every second day for the first week, and then each week

until the hardening is complete. In order to ascertain when this stage has been reached, the specimens should be removed from the jars and gently pressed with the fingers, when if the desired result has been obtained, they will be found to be tough but not brittle. They are next placed in water, which should be frequently renewed until the washings are colourless, and are finally put into methylated spirit until an opportunity is found for cutting them.

Bichromate of potash.—This salt is usually employed in the strength of from two to five per cent. watery solution, commencing with the former and gradually increasing the amount of the potash salt until the latter strength is obtained. It is used for the same class of specimens as Müller's fluid, but its action is rather quicker ; three weeks being required for specimens of ordinary size, but six weeks for a spinal cord, and six months for an entire brain.

The after-treatment is the same with both reagents.

Another form in which bichromate of potash is used is Erlicki's solution.

The composition of this is as follows :—

    Bichromate of potash . . .   2·5 parts,
    Sulphate of copper   1   part,
    Water 100   parts.

This reagent owing to the presence of the sulphate of copper hardens very much more rapidly than the previously described chrome preparations. A spinal cord may be hardened in it in ten days at ordinary temperatures, or in four days at a temperature of about 100° F.

To give a complete list of chromic acid salts which are used for hardening purposes the following must be mentioned.

Chromic acid.—This hardens extremely rapidly,

most specimens being ready in a few days. The
strength is gradually increased; at first it should not
exceed ⅓ per cent., and this should be gradually raised
to ½ per cent. It is very important that small pieces
of tissue should be placed in it, and large quantities of
the solution employed. The fluid must be changed at
the end of the first, second and third days, and then
every third day until the hardening is complete, this
being tested at frequent intervals as already described,
for if allowed to remain too long the specimens become
brittle and almost useless. This may in part be prevented by adding a little glycerine to the reagent. Care
must be taken to thoroughly wash out the acid with
water before placing them in alcohol for preservation.

Chromic acid and spirit.—This mixture, first
introduced by Urban Pritchard, is occasionally employed instead of chromic acid. It is made thus :—

Chromic acid . . . . . 1 part,
Water . . . . 20 parts,
Rectified spirit . . . . . 180 parts.

The chromic acid must be dissolved in the water, and
then the spirit added, otherwise a too violent action will
ensue.

This mixture is especially suited for hardening specimens of the eye or internal ear.

Ammonium chromate and **bichromate of
ammonia.**—Both these salts are occasionally used
as hardening reagents, and in a similar manner. They
are suitable for all tissues, but the former is especially
useful for preparing the kidney and other secretory
glands. The strength of each is the same, *i.e.*, five per
cent. The portions of tissues should be very small, and
placed in about fifteen times their bulk of the solution.
With ammonium chromate the process is complete in

forty-eight hours, and the specimens are then thoroughly washed in water, and either cut at once or removed to spirit for preservation. With bichromate of ammonia the time occupied is much longer. The fluid must be changed at the end of the first, third and seventh days, and at the end of each week until the sixth, when the tissue will be ready for cutting.

It will be seen from the above descriptions that the chrome salts usually require a long time before the hardening process is complete, and as they penetrate slowly, this depends in a great measure upon the size of the specimens. Frequent changes are necessary and therefore much more attention is required than when alcohol is used. Various alterations are also produced by their action in the interstices of the tissues. A peculiar network due to coagulation of albuminous substances very often takes place which may lead to errors if not rightly understood. Another disadvantage is that if the pieces are allowed to remain too long in the fluid they will be over-hardened and become brittle.

Corrosive sublimate.—This, though seldom used, is a capital hardening reagent. It is especially useful for glandular structures. The best strength is a half-saturated alcoholic solution, this is made by dissolving in one portion as much of the salt as 70 per cent. alcohol will take up, and then adding to it an equal quantity of alcohol of the same strength. The specimens must not remain in it too long as it hardens with great rapidity. A quarter of an hour to two hours (depending on the size of the specimens) is quite sufficient. At the end of the process the corrosive sublimate must be washed out with alcohol, otherwise the sections will be marred by the presence of black spots.

Platinum chloride.—This reagent is particularly

suitable for delicate objects such as the retina. It is usually made up in what is known as Merbrel's solution. This consists of equal volumes of 1·4 solution of chromic acid, and 1·4 solution of platinum chloride. Its action is fairly rapid, the process being complete in from three to four days. No change is necessary, and the reagent should be washed out with dilute alcohol (60 per cent.).

Osmic acid—Owing to its expense, and the irritating property of its vapour, this substance is rarely used for hardening purposes, except in the Pal-Exner method of staining the central nervous system, and for portions of retina. It is sold as a 1 per cent. solution, and must be kept in opaque bottles in the dark. Specimens should be submitted to its action for about ten hours, and then thoroughly washed in water. One of the most important uses of osmic acid is the selective power which it has on fat cells, these becoming blackened. A good plan in demonstrating fatty changes is first to harden the specimens in Müller's fluid and then to place them for a few hours in osmic acid. Another point which may be mentioned here is that turpentine and some other essential oils will decolourise particles of fat blackened by this reagent, and that therefore when sections have been cut and stained, the process of "clearing" by these oils must be rapidly conducted.

Picric acid.—A saturated solution may be used. The albuminates are gradually transformed into the insoluble form, so that hardening occurs without any shrinking. At the same time the specimens are stained a deep yellow which is very advantageous as a counter-stain to carmine. It is especially useful for epithelial structures, intestine, and the various tumours. After a stay of forty-eight hours in the acid, the specimens are washed in water, and preserved in alcohol. Picric acid

is occasionally employed as Kleinenberg's solution. This
is prepared as follows :—

Saturated watery solution of picric acid . 100 parts,
Strong sulphuric acid . . . . 2 ,,
Filter and add distilled water . . . . 300 ,,

This hardens more quickly than picric acid alone
(three to twelve hours) and is especially adapted for the
rapid hardening of various soft structures such as sarcoma or myxoma. The sections should afterwards be
stained in carmine.

DECALCIFYING.

In order to prepare sections of bone or of tissues in
which lime salts are deposited it is necessary to remove
these in order that the substances may be cut with a
razor, for the process of grinding down such specimens
until they are thin enough to be properly mounted is far
too tedious, and the results which are produced are
unsatisfactory.

The process of decalcification is carried out in much
the same way as the process of hardening, except that
of course the fluids employed are not quite the same.
Only the principal reagents will be described here.

Nitric acid.—This is used in two strengths, for
large bones one part of pure nitric acid is diluted with
ten parts of water, whilst for smaller bones and for
specimens containing calcareous deposits, this must be
further diluted to one per cent.

Specimens are first treated with absolute alcohol for
three days, and are then placed in the dilute nitric acid
for eight or ten days, the fluid being changed daily.
They must be frequently examined to ascertain if the

process of decalcification is complete, this being tested by piercing them with a needle. When ready they are washed for some hours in water and then placed in alcohol for preservation. A modification of this process, and a very excellent one consists in making a mixture of equal parts of 3 per cent. of nitric acid, and 70 per cent. alcohol. This acts more slowly than the previous one, requiring several days or even weeks.

Perenzi's solution.—This is perhaps about the best decalcifying fluid. Its composition is as follows:—

½ per cent. chromic acid . . . . 3 parts,
Absolute alcohol . . . . 3 ,,
10 per cent. nitric acid . . . . . 4 ,,

Its action is very rapid, the specimens being ready in one week. The after-treatment is the same as for nitric acid alone. The advantage of the three fluids just mentioned over other decalcifying fluids is that no swelling is caused, whilst at the same time no injury is suffered by the tissue elements.

Hydrochloric acid.—The strength of this acid for decalcifying purposes is 10 per cent. The process is much slower than with nitric acid, requiring a month before it is completed, in addition the acid causes the soft tissues to swell, so that their structure becomes almost unrecognisable. To avoid this a little chromic acid or alcohol may be added.

Von Ebner's solution.—This is a modification of the preceding, and has the following composition :—

Hydrochloric acid . . . .   5 parts,
Alcohol . . . . 1000 ,,
Distilled water . . . . . .   100 ,,
Chloride of sodium . . . .   5 ,,

A large quantity of the fluid must always be used and frequently changed, the specimens being also occasionally tested to ascertain if they are soft enough for cutting ; they are afterwards thoroughly washed in water.   If it be desired to examine the fibrillar structure of bone, the sections should be mounted in 10 per cent. salt solution, but if this is not a matter of importance, the ordinary mounting media may be employed.

Picric acid.—A saturated watery solution of picric acid may be used, but is not altogether satisfactory. Its action is weak and very slow, in addition it causes considerable shrinking of the tissues.   A large volume of the reagent must be employed, and a few crystals of the acid should always be kept at the bottom of the jar to ensure the strength being maintained.   When decalcification is accomplished, after washing in water, the specimens are preserved in spirit.   This process is only suitable for very small pieces of young bone.

CHAPTER III.

EMBEDDING.

IN order that thin and uniform sections may be made it is not sufficient that the specimens have been hardened or decalcified. If it be attempted to cut the specimens without further preparation, it will be found that the sections easily break directly they are made, if indeed it be possible to obtain any at all. In addition to these objections it is very difficult to hold the specimens firmly enough without crushing them and greatly interfering with their structure.

In order to overcome these obstacles, it is necessary either to surround the material with some substance firm enough to afford a hold to the hand, or to saturate it with a medium which will impart to it a uniform consistency. This operation is known as "embedding."

From the above prefatory remarks it will be seen that embedding is necessarily of two kinds, viz., "simple" embedding, which is fixing or placing the object to be cut in a medium firm enough to hold it, and "interstitial" or "infiltration" methods, which consist in permeating the specimens with some medium, which is at first fluid, but gradually becomes solid; this being accomplished by the use of substances which are liquid when hot and solid when cool, such as paraffin, or by dissolving some substance in a volatile fluid, such as celloidin in ether and alcohol. A third method, and the

one which is perhaps the most commonly adopted, consists in saturating the specimens with a solution of gum,
which is afterwards frozen.

The **freezing media** will be first described. The
initial stage consists in abstracting the hardening reagent. This is of course absolutely essential, as alcohol
will not freeze, and the chrome salts would be deposited
irregularly.

The specimens therefore are placed in a basin containing ordinary tap water which should be frequently
renewed. The process is shown to be complete by the
specimens sinking to the bottom of the containing vessel
if they have been hardened in alcohol; or by the water
remaining untinged if chromic acid or its salts have been
used.

They are now ready to be placed in gum. The solution usually employed is gum and syrup, which is made
of equal parts of gum solution (B.P. strength) and syrup,
to which a little thymol should be added as a preservative.

The following, however, recommended by Mr. T. L.
Webb ("The Microscopic Journal," ix., 1889, page 344),
will be found preferable. It freezes sufficiently to give
a firm support without being too hard, it keeps better
than gum, and is in addition cheaper.

An aqueous solution of carbolic acid (1 in 40) is taken
and sufficient dextrin is dissolved in it to make a thick
syrup.

Dextrin dissolves slowly in cold water, so that a
gentle heat must be used when making the mucilage.
It solidifies if kept for some time, but may be liquefied
again by gently warming the bottle.

The specimens should be allowed to remain in these
solutions for twenty-four hours, so as to become thoroughly permeated.

Simple Embedding.

Paraffin is usually employed for this method.
A little tray is first made out of light cardboard in the
following manner (fig. 19). The card should be oblong
in shape as in the diagram. It is first folded along the
lines AA' and BB', next along the lines CC' and DD',
keeping the folds of the paper on the same side. The
lines aa', bb', cc' and dd' are then marked with a pencil.
The corners are now pinched up between the finger and

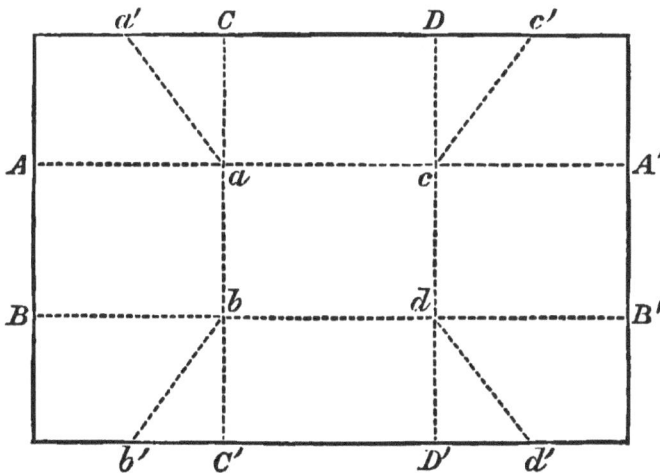

Fig. 19.—Diagram for embedding box.

thumb and bent round so as to lie against the sides of
the tray. The sides are secured by doubling tightly
down the projecting corners. Two paraffins are usually
employed, a hard one melting at 49° C., and a soft one
melting at 46° C. Two parts of the former are mixed
with one of the latter, the resulting mass yielding a good
consistency for cutting. If it be found too hard, a little
vaseline may be added to soften it.

E

A small piece of the paraffin mixture is heated over a water bath until the paraffin melts, and it is then poured into a cardboard tray. The specimen meanwhile must have been placed in alcohol, and left there for a few hours. It is then removed and dried as far as possible by means of filter paper. As soon as the melted paraffin in the tray begins to set at the edges, the specimen is immersed in it and the mass allowed to solidify around it. When cool, a block of paraffin is cut out of the box so as to contain the specimen and afford a convenient support whilst sections are being cut.

Infiltration Methods.

In medical work, " simple embedding " is very seldom employed, it gives almost as much trouble as the methods now to be described, and the results are not nearly so good.

The two chief media employed for the infiltration methods are celloidin and paraffin. In England the latter is more usually adopted, whilst abroad, and especially in Vienna, celloidin is generally chosen. It is difficult to state which is the better of the two, but the celloidin method is more easy to learn, and the results are all that can be desired. It affords an excellent consistence for cutting, and at the same time is so transparent and flexible that it may be mounted with the specimen without detracting from the microscopical appearances; its only drawback is, that it is apt to retain some of the stains, especially the aniline dyes, but even this may be obviated with a little care. If sufficient trouble be taken in the preparation of the solution, sections may be cut as thin as with paraffin, although it is generally stated that, when very thin sections are required, paraffin *must* be used. If both these methods are employed, a very good rule is celloidin for large, paraffin for small specimens.

Embedding in celloidin.—Celloidin is a preparation of pure pyroxylin. It can be obtained in two forms, in tablets or in cuttings; the latter being far more convenient. The substance manufactured by Schering is the best.

To prepare it for use, a few "cuttings" are placed in a small wide-necked bottle; a mixture of absolute alcohol and ether, equal parts, is then poured in and allowed to remain for some hours, as the celloidin dissolves very slowly; it should be stirred with a glass rod from time to time, and sufficient of the mixed fluids added to form a mixture of about the consistence of thin mucilage.

When dealing with very delicate specimens, such as spinal cord, the eye, embryos, &c., it will be found a good plan to prepare two solutions, one very fluid, and the other rather thicker, the specimens being placed first in the thin and then in the thick.

If the specimens have been hardened in Müller's fluid or any of the chrome preparations, this must be removed by thorough washing in water before they are removed to dilute spirit and then to absolute alcohol. If on the other hand they have been hardened in alcohol, the process of embedding may be at once proceeded with. The first stage consists in soaking the specimens in a mixture of equal parts of absolute alcohol and ether; in this they must remain for about twelve hours. They are then transferred to the celloidin mixture, if necessary first to the thin and then to the thicker solution. Here

again at least twelve hours stay is necessary. Small
blocks of cork are then prepared; an ordinary bung cut
into six pieces answers very well. It is important that
one surface of the cork should be quite flat and perfectly
dry. A large glass jar with a glass stopper will also be
required; this must be about half filled with 80 per cent.
alcohol.

After the specimens have been a sufficient length of
time in the celloidin, they are removed one by one by
means of forceps and each placed on a cork; a few
minutes are allowed to elapse for the celloidin to begin
to set, and the corks are then thrown into the jar containing the alcohol. With a little manœuvring, they
can be arranged without contrivances to sink the corks,
the weight of the specimens being sufficient for this
purpose.

They are allowed to remain in the alcohol until the
sections can be cut, but a minimum time of twelve hours
is necessary.

Embedding in paraffin.—As in the case of celloidin the first stage of embedding is the saturation
of objects with some solvent of the embedding material.
There are various liquids recommended, turpentine,
clove-oil, benzine, chloroform, xylol or cedar-oil. The
latter has been highly praised as penetrating rapidly,
mixing readily with the paraffin and not tending to
render the tissues brittle. Turpentine, however, is
most commonly used, and except for very delicate structures, when cedar-oil is to be recommended, answers
every purpose.

The specimens are first dehydrated by means of
absolute alcohol, in which they should remain for
several hours.

The specimens are placed directly from the alcohol

into the turpentine or cedar-oil, and will be completely cleared in from one to six hours according to their size.

A mixture is then made of two parts of hard paraffin (melting at 49° C.), and one part of soft (melting at 46° C.), the mixture being placed in a small copper vessel or porcelain dish, and the paraffin melted and retained at a temperature of about 1° above its melting point; it is a matter of great importance not to heat the paraffin too much, but to keep it as nearly as possible at its melting point. It is this difficulty that renders celloidin preferable to paraffin.

The objects are removed from the clearing mixture (turpentine, &c.), and immersed in the melted paraffin where they should remain from three to five hours, when they will be found to be completely saturated.

The final stage in the paraffin process is the " embedding " proper. Small trays made in the way described above, are filled with melted paraffin, and the objects are removed from the warm paraffin in which they have been lying and immersed in the liquid mass, and the whole should be cooled as quickly as possible by placing the trays in running water, but not permitting them to be completely submerged. If the paraffin be allowed to cool slowly, there is great danger of its crystallising, and so not forming a homogenous mass.

Instead of the trays, embedding L's may be used. These consist of two L-shaped pieces of lead, about half-an-inch high, the long arm being about two inches long, and the short one three quarters of an inch. A glass slide is coated with glycerine, and the two pieces of lead laid on it so as to form a rectangular box, which is then filled with the melted paraffin.

An extremely simple process, which may be adopted when the specimens are small, is to pour the melted mixed paraffins into a watch-glass or small porcelain cup, and then to place the specimens after soaking in turpentine or chloroform in the melted wax. The watch-glass or cup is then placed in an oven maintained at a temperature of about 52° (so that the paraffin is kept at about its melting point) for five or six hours. They are then removed and allowed to cool. When the embedding material has set, the glass is slightly warmed so as to loosen the wax, or the specimen may be cut out with a knife and the block pared, as described in the next chapter, ready for the microtome.

CHAPTER IV.

CUTTING SECTIONS.

In the previous chapter the methods of hardening and embedding specimens have been described, and the next process is cutting the sections.

For reasons already stated, at the present time sections are very rarely cut free hand, but generally by means of a microtome. The principal instruments have been described in Chapter I., it now remains to describe their use and their application to frozen and embedded objects.

Except for special purposes, the method by freezing is the one generally adopted in this country; it has the advantage of being very speedy, and by practice almost as good sections may be obtained by this means as by the more complicated processes.

It is hardly applicable, however, to delicate specimens, such as the eye, and for bacteriological purposes it cannot be recommended, for the reason that when the gum has been frozen and melts again small particles are very apt to fall out. Take for example, sections of tubercular lung. Excellent sections can be obtained by freezing the more solid parts where the fibrous tissue is in excess, but when breaking down of tissue has been going on, or when small caseous masses are present, it is extremely difficult to get good sections owing to the fragile nature of the tissues and the caseous nodules will probably be lost; consequently, the search for tubercle bacilli is greatly interfered with, and this drawback is added to by the multiplicity of fluids through which the sections will have to pass before the staining is complete ; for such specimens, therefore, it is advisable to embed in celloidin. In many of the continental laboratories, freezing is but rarely resorted to, nearly all the specimens being embedded in celloidin, and the sections obtained are most valuable, far excelling in most respects the ordinary sections obtained by freezing, although as before stated, by dint of long practice, very good sections can be thus prepared.

For certain purposes it is often desirable to cut sections in "series," that is to say, to obtain a number of sections in the order in which they are cut, without losing any of the intermediate ones. It is possible to do this with the ordinary microtomes, but not to nearly such perfection as can be done by the Cambridge Rocking Microtome. "Series" cutting, however, will be separately considered.

We now proceed to describe in detail the methods of cutting sections, and we will first deal with frozen specimens.

It will be presumed that all the foregoing processes for hardening, &c., have been duly carried out, and the objects which are to be cut placed in the gum.

Cutting with Cathcart's Microtome.—One of the specimens is removed from the gum by means of forceps, and placed on the plate of the microtome. A little of the mucilage being added so as to completely envelope it. The spray points having been set in position, the ether bottle is filled with ordinary methylated ether. The spray apparatus is then set to work, rather briskly at first, until a white ring forms at the edge of the gum, which gradually spreads until the whole is a frozen mass firmly fixed to the plate. After this the bellows must be worked more gently. Throughout the whole time care must be taken to brush off the frozen vapour which in a moist atmosphere may collect below the plate, and if not removed will interfere considerably with the freezing action of the ether. So much so indeed that the specimen may become loosened. If the ether should tend to collect in drops below the plate, the bellows must be worked more slowly. Occasionally the ether points become choked, they must then be cleared by means of a piece of fine wire.

When the object is completely frozen, the left hand is used to turn the milled head, whilst the right hand grasps the plane-like knife. The plate is first lowered as far as possible, and then the knife, at about an angle of forty-five degrees, is passed slowly over it, so as to cut off any portion of the specimens projecting above the level of the glass slide. The micrometer screw is then slightly turned so as to raise the specimen, and the knife is again passed across it. A section will thus be cut. Practice alone can regulate the thickness of the section. In his latest instruments Mr. Cathcart has so arranged that when the screw is turned through the eighth of an inch, the plate rises one two-thousandths of an inch.

The above operation takes a long time to describe but only a few moments to perform. By rapidly repeating it some dozens of excellent sections may be obtained in the course of a few minutes.

Several sections may be allowed to collect on the knife before removing them. They should be detached by gently sweeping the finger across the knife or by means of a camel's hair brush, and should be immediately transferred to a shallow dish containing water when by cautiously manipulating them with a needle they are easily separated from one another.

Before again proceeding to use the knife a few puffs should be given to the spray apparatus to ensure the mass being sufficiently frozen. It is sometimes advised that the knife should be moistened with water, but this is unnecessary as the thawing of the specimen yields quite sufficient moisture to prevent the sections breaking.

The knife should not be carried straight across the specimen, but rather obliquely, so as to expose a greater portion of its edge for cutting purposes, and it should be passed quickly and evenly along the guides. After use the knife must be dried with a soft cloth, and passed a few times over a sharpening stone, especial care being used to remove all notches, and secure a straight, sharp and smooth edge.

It is almost needless to add that the microtome must be firmly clamped to the table and so arranged that the plate is perfectly horizontal.

Cutting by Swift's microtome.—The preliminary operation of freezing is carried out in precisely the same manner as just described. The cutting apparatus, however, is quite different. It consists of a tripod stand supporting a razor, and is known as a "plough." A minute description of this instrument has already been given in Chapter I. When arranging it for use care must be taken to see that the two posterior screws are on the same level, and the edge of the razor must be raised or lowered by means of the front screw until its edge is on a level with the top of the specimen. Thus, whereas in Cathcart's microtome the specimen is raised to meet the razor, in Swift's instrument the razor is lowered to meet the specimen.

When the object is sufficiently frozen, the plough is grasped with the two hands, the forefinger of the right being extended so as to turn the large head of the front screw. The razor is then passed obliquely across the specimen, and then drawn back again, the three supports of the tripod being firmly pressed on the glass plate. The front screw is next slightly turned, and the process repeated, the thickness of the sections being thus regulated by means of this micrometer screw. When five or six have been cut they are cast into water in the same way as has been previously described.

It is a good plan to moisten the glass plate of the microtome with a little glycerine, so as to enable the tripod to run easily. After use the knife and the entire apparatus should be carefully dried.

The use of Williams' ice-freezing microtome.—This instrument is too cumbersome for ordinary use, it takes a long time to prepare, and has no advantage over the two already described, except when a very large number of sections have to be prepared for class teaching.

The ice-box is filled with salt and ice in alternate layers until the box is full. The lid is then fastened on and one of the discs screwed into place. The specimen is removed from the gum and placed on the disc, a little excess of the embedding fluid being added. A few minutes are allowed to elapse in order that the gum may freeze, and sections are cut with Swift's plough in precisely the same manner as when using his microtome.

Cutting sections embedded in celloidin.—For this purpose any of the microtomes may be used. If Cathcart's be employed the disc must be removed, and the clamp fixed in its place. The same if Swift's be chosen. It is possible to combine the freezing and celloidin methods ; this is accomplished by cutting the specimens away from the corks upon which they have been fixed (see Chapter III.), and placing them on a disc of a Cathcart's or Swift's microtome, surrounding them with water, and fixing them by freezing.

The results, however, are only fairly satisfactory, and the process cannot altogether be recommended. The chief drawback is that alcohol cannot be used for soften- ing the embedding material, and consequently the sections are extremely apt to break and curl up.

The best instrument to use, however, for cutting specimens embedded in celloidin, is **Reichert's mi- crotome.** Owing to the steadiness of the stand and the mechanical arrangement of the razor, most excellent sections can be obtained by its employment. The objects having been duly hardened, embedded and fixed on corks as described in the previous chapter, are kept in the eighty per cent. alcohol until required. One of them is now removed and the cork fixed in the clamp of the machine, and so arranged by means of the screws that its upper surface is just below the level of the razor. The angle at which the specimen is set can only be determined by experiment, but it is a very important point ; if the shape of the object is triangular, it is preferable that the razor should first encounter its largest side ; if it is quadrilateral, one of the angles should be placed innermost, but any projecting corner should always point outwards, otherwise a complete section will rarely be obtained.

The specimen having been satisfactorily set in posi- tion, the next step is to arrange the knife. The rule is to allow as much of the cutting edge of the knife as possible to come into action, that is to say, the razor should be set very obliquely, so that it may be drawn across the tissue from heel to point.

Before using the razor, care must be taken to see that its edge is perfectly smooth and sharp, and also that its carriage works easily and smoothly.

Probably the surface of the specimen will at first be uneven, and will not occupy the whole extent of the tissue, a few preliminary cuts must therefore first be made, so as to expose as large a surface as possible. The specimen may be raised between each cut in two ways, either by moving the micrometer screw with the finger, or by the automatic apparatus, by means of which the screw is turned by a stroke of the razor at the termination of each cut. If the specimen be an exceptionally good one as regards consistency and uniformity of structure, the self-regulating arrangement is an excellent one, but this unfortunately is seldom the case, and it is therefore better to practise the first method, so that the thickness of each section can be accurately regulated.

If sections can be obtained when three or four notches of the wheel have been turned, these being indicated by audible "clicks," the operator may be assured that he has succeeded in cutting beautifully thin sections; if five or six clicks have to be given before a complete section results, it will be found thin enough for ordinary work, but beyond this, thick specimens only will be made, and some misfortune (probably unavoidable) has occurred in the previous treatment of the tissue, either it is of too diverse a structure to yield large sections (*e.g.*, eyes or embryos), or it was not placed early enough in the hardening fluid, so that decomposition had set in ; or the morbid change is of such a nature that it is extremely difficult to obtain sections at all, such as often happens with specimens of myelitis, or with caseous material in various organs. The correction of some of these difficulties will be considered immediately.

In cutting celloidin specimens it is of the utmost importance that the knife and tissue should be kept covered with dilute alcohol (50 per cent.). The spirit should be placed in a capsule immediately beneath the razor, and by means of a camel's-hair brush fre- quently swilled over the razor and celloidin ; if this be not done, the specimens will tend to break or crumble before the knife, or curl up in such a manner that it will be found almost impossible to unroll them.

Sometimes, however, in spite of all these precautions, it will be found very difficult to cut delicate specimens, and various artifices have to be adopted.

˙ In the first place, sections may sometimes be obtained when other efforts have failed, by drawing the knife very slowly over the specimen, at the same time gently unfolding the section by means of the brush, and spreading it out on the razor. Changing the direction of the specimen by removing the cork from the clamp and turning it round, so that another part of the tissue comes first into contact with the knife, will occasion- ally produce the desired result. But the most suc- cessful plan is to employ what is sometimes known as **"collodionisation of the sections."**

This is accomplished as follows (" Quart. Journ. Roy. Mic. Soc.," 1885, p. 738). Some very fluid collodion is placed in a small bottle ; it is convenient to have a bottle with a perforated cork, through which passes a camel-hair brush, or one with a glass stopper to which a small rod is attached. The collodion must be ren- dered so thin with ether, that when applied to the surface of the celloidin it dries in a few seconds, and must be further diluted as soon as it begins to leave a shiny surface on the tissue.

In dealing with very brittle objects, the moistened brush or rod (it must be no more than merely moist) is passed quickly over the surface of the tissue, so as to leave a thin film ; as soon as this is dry the section is cut. This method is especially useful in dealing with the central nervous system. It is not well to allow too many sections to accumulate on the knife, but after a few have been cut, they should be transferred by means of a camel-hair brush to a shallow dish containing water ; owing to the rapid separation of the spirit they immediately spread out, and then may be transferred to dilute spirit preparatory to mounting.

When the sections are very fragile, it is better to place dilute spirit (50 per cent.) in the dish instead of water, otherwise they are very often torn in pieces by the rapid way in which the mixing of the spirit and water causes them to unfold.

Cutting sections embedded in paraffin.— The paraffin blocks containing the objects are pared so as to fit the clamp, which is arranged in the same manner as already described for the celloidin specimens. The succeeding processes, however, are quite different. The sections must be cut dry, that is to say, no fluid must be used to moisten the knife. The razor also should be set as square as possible, for if the oblique position be adopted, there is a great tendency for the sections to curl up. Should this accident occur they may sometimes be made to unroll by placing them in water at a temperature of 40° C. After they are cut the sections should be placed in turpentine or xylol, in order to remove the paraffin. This takes some time, but after it is successfully accomplished they must be passed through absolute alcohol and weak spirit to water, before commencing the ordinary methods of staining and mounting.

Cutting in series.—For medical work, the process of cutting a number of specimens in succession, so that a whole series may be preserved, is almost limited to the examination of the central nervous system and the pathological examination of the eyeball. In the physiological laboratory it is more commonly employed.

Undoubtedly the most beautiful and complete series can be obtained by means of the Cambridge rocking microtome, but for ordinary purposes a very good series can be prepared by Weigert's method with an ordinary microtome ; this is carried out as follows :—

The specimens are best embedded in celloidin. Sections are cut in the usual manner, but only one must be on the knife at a time and should be carefully spread out with a camel's-hair brush. A thin strip of tissue paper a little wider than the section is laid upon it, to which it adheres, and may thus be removed. In this manner each section is taken up in the order it is cut, as many strips being made as is necessary for completion of the series. Each strip should not be more than three inches long, and when filled should be laid on filter paper moistened with dilute spirit so as to keep it damp. Glass slides are now coated with a thin layer of collodion and allowed to dry. On these the strips of paper are placed prepared face downwards, so that the sections come into contact with the collodion. A little pressure with dry filter paper will cause them to adhere firmly, so that the paper may be removed. Some more filter paper is then gently pressed over the sections to dry them, and finally a second layer of thin collodion is painted over them, and they are thus preserved sand-wich wise between the two layers. When the slides are immersed subsequently in the various staining fluids the double film is usually set free, carrying the sections with it. Each film can then be treated as a section, and after the staining process is finished may be cleared in xylol, and again fixed to a glass slide by means of collodion. Instead of a cover glass a layer of photo-grapher's varnish may be applied. In the preliminary stages, instead of painting the collodion on to the slides with a brush, it is better to pour a little collo-dion on the slide, and then by means of tilting, a thin uniform layer may be obtained in the same way as in the preparation of photographic plates.

Cutting with the rocking microtome.—The object will probably have been embedded in paraffin ; if celloidin has been employed, the specimen must be removed from the cork, and after the edges have been carefully pared so as to just expose its surface, it is immersed for a few minutes in benzol, and then covered with paraffin wax by the method described on p. 49.

The specimen is fused on to the paraffin contained in the socket of the microtome by means of a heated knife, being afterwards trimmed, so that a narrow border extends beyond the object in all directions.

Dr. Lovell Gulland ("Journal of Anatomy and Physiology," vol. xxv., p. 56), gives many useful hints on this subject. He lays stress on the importance of making the surface which meets the razor exactly parallel to the opposite surface, and of the whole block being exactly rectangular. A thin layer of soft paraffin is next applied to the upper surface (the one meeting the razor), and to the opposite surface ; this is best carried out by dipping these surfaces into the melted

F

soft paraffin, and when this has become firm, the surfaces are again trimmed square.

This is done in order to prevent any curve in the
ribbon from being accentuated by the flattening out of
the sections.

The specimen mounted on the lever is brought near
the razor by turning the milled head, and so raising the
horizontal arm. The ratchet in connection with the
handle is most conveniently arranged so as to catch
seven teeth at each stroke, corresponding to about $\frac{1}{5000}$
of an inch thickness of the section. The handle of the
microtome is worked backwards and forwards with the
right hand, whilst the left manipulates the ribbon of
sections as it falls.

Dr. Gulland advises that the ribbon should be
divided into lengths corresponding to that of the slide
in use. These divisions are to be seized at one end
with forceps, and the other end is gently lowered on to
the surface of warm water contained in a flat dish,
placed just beneath the microtome ; as the sections
flatten out they will be found to move along the top
of the water so that more and more of the ribbon can be
lowered. When the flattening is complete, the slide
carefully cleaned is immersed in water, and the ribbon
floated into position on the slide with a stiff brush,
this process being repeated until the slide is full, the
ribbons being laid side by side. It is then set up on
end to allow the water to drain off. The temperature
of the water must be regulated according to the melting
point of the paraffin, but it should never be hot enough
to melt the wax ; short of this, the warmer the water
the more rapidly and completely are the sections flattened.

The sections are fixed on the slide by evaporating

the water from its surface, preferably over a warm oven, just warm enough to drive away the moisture without melting the wax. Thus, if the paraffin melts at 52° C., the oven should be 50° C.

When the water has completely evaporated, the opacity of the sections disappears, they look dry and become much more transparent. When the fixation is complete, the paraffin is melted and washed off with turpentine or xylol.

One of the greatest advantages of this method is the perfect ease and safety with which it allows sections to be manipulated on the slide, so that the most varied stains and reagents may be directly applied.

CHAPTER V.

STAINING.

SECTIONS may be examined directly they are cut without undergoing the process of staining; although considerable information may be gained in this way, far more can be obtained when certain dyes have been employed; in this way special structures are brought out which would otherwise have escaped notice, and the various tissues are differentiated one from another owing to their unequal receptivity for the stain or stains.

The chief object of staining is therefore a selective one. The differentiation may be accomplished either by the employment of one stain, which picks out certain elements of the tissue, either leaving the rest colourless, or by a chemical action producing quite a different hue. An example of the first class is hæmatoxylin, which picks out the nuclei, leaving the ground substance almost uncoloured, or only of a delicate violet tint; an instance of the second class is methyl aniline violet in presence of amyloid material; this it stains red, the rest of the section assuming a violet tinge. A more common plan is to use two stains, one of which colours the nuclei, and the other the ground substance; for example, hæmatoxylin and eosin, carmine and picric acid.

The selective power of stains is aptly classified by Mr. Arthur Lee (" Microtomist's Vade Mecum ") into " histological selection and cytological selection." In the former, an entire tissue or group of tissue-elements is prominently stained, the elements of other sorts present in the preparation remaining colourless, or being differently stained, as in a successful impregnation of nerve endings by means of gold chloride. In the latter, the stain seizes on one of the constituent elements of cells in general, either the nucleus, or the extra-nuclear parts.

Owing to the selective power of the different stains, they have been divided into nuclear, general and selective, according to whether they colour the nuclei only, the whole tissue, or particular elements such as bacteria and fat (see " Manual of Clinical and Practical Pathology," by Dr. Wynter and the Author).

There are two chief methods of staining, the **direct** and the **indirect.** In the first, the tissue is coloured by some dye which seizes upon one particular group of elements, such as the nuclei, more quickly than upon the other constituents of the tissue, and the process is stopped when the former are sufficiently coloured, whilst the latter are merely tinged ; an instance of this class is hæmatoxylin.

The indirect method, however, applies to the great majority of stains. In this, the tissues are first overstained, and then the colour again partially removed by some agent, certain elements retaining the colour longer than others. An explanation of this action is difficult to find ; in the case of bacteria, an ensheathing envelope may be the resisting power to the decolourising media. The aniline dyes belong to this class, and most beautiful results are obtained by their use.

In this chapter the use of those stains will be considered which are applicable to ordinary histological processes ; a description of the selective stains, such as those suitable for the central nervous system, and those used for bacteriological purposes will be given in succeeding chapters.

A few words. are necessary concerning the management of stains generally, and the manipulation of sections.

The stains should be procured as pure as possible, and therefore it is much better to get them from the regular dealers, such as Becker, Pillischer, Martindale, Allan and Hanbury, &c., than from any chance druggist. They are best preserved in smoked glass bottles.

All stains must invariably be filtered before use, as deposits are very apt to occur, and unless these be separated, the sections will be entirely spoiled.

For holding the stains when in use, small glass capsules or watch glasses with ground under-surfaces to keep them steady are usually employed, but the most convenient are solid blocks of glass hollowed out, such as are sometimes used as salt cellars. These hold about the right quantity of fluid, are of the right depth, very firm, and can with safety be covered over with small squares of glass so as to exclude dust. This is most important, especially when the sections have to remain for several hours exposed to the colouring agent, and is absolutely essential when alcoholic solutions are used.

It will be seen that some of the dyes are in watery solutions, others in alcoholic ; when they have not been diluted, they may be filtered back into the bottle after use for the sake of economy, but the alcoholic solutions have frequently to be diluted, and they cannot then be employed a second time.

In the preceding chapter, after the sections were cut, they were directed to be kept in dilute methylated spirit (equal quantities of spirit and water) until wanted. If the dye has been dissolved in spirit, the sections may be transferred directly to the stain, but if a watery solution be employed, they must be rinsed for some minutes in water until the alcohol is removed, and with delicate sections care must be taken that the rapid separation of the spirit does not cause them to fly into pieces.

A section is most conveniently transferred from one solution to another on the end of a glass rod drawn out to a fine point ; a gentle movement of the rod will dislodge the section, and with a little practice it may be easily made to spread out. In the early stages it is not essential that the section should be kept flat, but it is advisable to do so if possible. Glass rods are to be preferred to needles, as they are less liable to catch in the section and tear it. The advice sometimes given to always move sections by means of a section-lifter is not altogether good, as a certain amount of pulling of the section is then required, and a delicate one will very likely be considerably damaged thereby. Special directions will be given for the manipulation of fragile specimens.

The choice of a stain.—In preparing ordinary sections (as distinguished more especially from bacteriological work) there are two combinations which are most commonly employed, namely, hæmatoxylin and eosin, and carmine and picric acid ; they yield excellent results, and the student cannot do better than employ them. A great number of other double and treble stains are constantly being suggested, but these two have stood the test of time. They are simple to carry out, not requiring a complicated series of changes from one reagent to another.

These methods will first be described.

Staining with hæmatoxylin and eosin.—
Sections stain best with hæmatoxylin, which have been
hardened in alcohol and Müller's fluid, but Delafield's
preparation will also stain preparations which have
been hardened in chromic or osmic acid.

The best formula is that which was originally introduced by Delafield.   It is made as follows :—

To 100 c.c. of a saturated solution of ammonia alum,
a solution of one gramme of hæmatoxylin, dissolved in
six c.c. of absolute alcohol, is added drop by drop.
The solution is exposed to the air and light, in an un-
stoppered bottle, for three or four days; it is then filtered, and twenty-five c.c. of glycerine and twenty-five
c.c. of methylic alcohol are added.   The mixture is
allowed to stand until its colour becomes dark, and is
then again filtered and preserved in a bottle with a
closely fitting stopper.   It keeps well, but should not be
used for two months after it has been prepared.

Another formula for making the stain is the one that
bears Ehrlich's name.   It can also be highly recommended, but it has no advantages over Delafield's, and
will not stain specimens which have been hardened in
chromic acid.   Its composition is as follows :—

Hæmatoxylin	2 parts.
Alcohol	100 ,,
Distilled water	100 ,,
Glycerine	100 ,,
Alum	2 ,,
Glacial acetic acid	4 ,,

Dissolve the hæmatoxylin in the alcohol, add the
glycerine and water and finally the acetic acid.   The
solution will not have acquired its full staining powers
until a week after it has been made.   The mode of
using the above solutions is the same.

There are two methods of staining by hæmatoxylin, a rapid and a slow. In the rapid process the sections are dipped for about half a minute into the undiluted stain and are then thoroughly washed in water. Speci-mens coloured in this way are not always satisfactory ; they are likely to be stained unequally ; being over-stained in one part and not sufficiently so in another.

The slow method yields far more satisfactory re-sults. Distilled water is placed in a capsule until it is about half full, a few drops of the stain are then added, and the whole filtered. The more dilute the mixture the longer must the sections remain in it. A convenient plan is to add about half a dozen drops of the hæmatoxylin solution, and then allow the specimens to remain in the mixture all night. They are then removed and thoroughly washed in water ; a stay of two hours or more in the water greatly improves the colour. After-wards one of the sections is placed on a glass slide, spread out and placed under the microscope, and ex-amined by an inch objective. It will then be seen whether it is properly coloured. If it be not sufficiently so, the specimens must be put back in the stain, but if they are too dark they may be washed for a few seconds n a five per cent. solution of acetic acid. Over-staining is to be avoided if possible, as after the action of the acid they do not keep their colour well and soon become useless.

After being properly stained in hæmatoxylin, a good effect is produced by " counter-staining " them in eosin.

For this purpose the following solution must be made :—

    Eosin   . , ...   5 parts.
  - Distilled water   100   ,,

The sections are removed from the water in which they have been washed and allowed to remain for about five minutes in the stain. They are then swilled in water and rapidly dehydrated in absolute alcohol, cleared in clove oil and mounted in balsam, as will be described in the chapter devoted to "mounting."

It must be stated here, that the eosin is quickly dissolved out of the sections by absolute alcohol and oil of cloves, and therefore these processes must be performed as rapidly as possible. It is a good plan to add a few drops of eosin to the alcohol.

Thus prepared the nuclei are stained a deep bluish-purple, those of the epithelial cells being less darkly coloured than those of the lymphoid cells and connective tissue. The ground substance assumes a clear rose colour, the fibrous tissue being especially picked out. This method is particularly applicable to specimens of cirrhosis.

Double staining by hæmatoxylin and eosin may be accomplished in one process by using Renaut's "hæmatoxylic eosin." It is made as follows:—

Concentrated watery solution of eosin . 30 parts.
Saturated solution of potash alum in glycerine . . . .} 130 „
Saturated solution of hæmatoxylin in alcohol . . . .} 40 „

The eosin solution is added drop by drop to the potash, and the mixture filtered; then the hæmatoxylin solution is also added drop by drop, and the mixture set aside for five or six weeks in a vessel with a perforated cover, until the alcohol is evaporated, when, after filtering, the stain is ready for use.

The best method for using this stain is to mount the objects in the fluid diluted with one or two volumes of glycerine, as the colour is very slowly absorbed. After a few weeks the specimen will have become perfectly clear. The cover glass should be fixed with balsam or gum damar.

This stain has a selective action on the cells of the salivary mucous glands; mucous cells it colours a pale blue, whilst the demilunes become a deep rose colour.

Staining by carmine and picric acid.—A very large number of preparations of carmine are used for microscopic purposes. As with hæmatoxylin and eosin a simple method will first be given in its entirety, and then the other preparations of carmine will be briefly considered.

Carmine is a most excellent nuclear stain. It is rather more trying to the eyes than hæmatoxylin, but is more permanent. The usual method for its use is first to stain the specimens throughout, and then wash out the excess of colour from everything except the nuclei. But if no such reagent be employed the carmine will be found to stain the nuclei a bright red, and in addition to colour, though not so brilliantly, the fibrillar ground-substance of the connective tissue, muscular fibres, the ground-substance of bony tissues, fibrine, the neuroglia of the central nervous system, the axis cylinders of nerves, &c., leaving uncoloured elastic tissue, the white substance of Schwann in the nerves, fat, mucous substances, chalky deposits, &c.

The most convenient formula is **lithium carmine.** This is very largely used in the laboratories of Vienna, but not so well known in England. It produces very good results, and its manipulation is easy to learn. It is made as follows :—

Carmine . . . .   2·5 parts

Saturated solution of lithium carbonate   100 parts.

The mixture requires to stand for some time and to be carefully filtered before use.

Sections are allowed to remain in the undiluted stain for about five minutes. A longer stay will not be harmful, as over-staining is not easily accomplished, in fact if the sections remain for twelve or twenty-four hours in the dye they will be found to be by no means spoilt.

After removal from the stain the sections are at once placed in **acidulated alcohol**:—

Hydrochloric acid . . . . .   1 part
Alcohol . . . .   70 parts
Distilled water . . . . . .   30 parts.

The specimens will be at once seen to give up a great deal of colour. They should occasionally be slightly moved about so as to distribute the cloud which forms round them. Two or three minutes usually suffices for decolourization; but they should not be removed from the solution so long as any colour comes out. It is often necessary to have two basins containing the dilute acid, so as to ensure all the superfluous colour being removed. They are next transferred to absolute alcohol to which five or six drops of a concentrated alcoholic solution of picric acid have been added. This dehydrates the sections, as well as counter-staining them, and after two minutes they may be removed to oil of cloves, or creosote, preparatory to mounting in balsam. As both these oils remove the picric acid rapidly, a few drops of the concentrated solution of picric acid should be added to the clearing reagent.

Staining with carmine and picric acid may be accomplished in one operation, by using the stain known as **" picro-carmine."** Sections coloured in this manner cannot be mounted in balsam. The solution is rather difficult to prepare, but the directions given by Dr.

Sims Woodhead (" Practical Pathology," p. 56) are thoroughly reliable, and the observer cannot do better than follow them closely.

" Take of—

Pure carmine . . . . 1 part
Strong ammonia . . . . 3 parts
Distilled water . . . . 3 parts.

Dissolve the carmine in a test tube with the ammonia and water. To this add 200 parts of a cold, saturated, and filtered solution of picric acid, and mix thoroughly. Place the fluid in a basin and cover with a clock glass (with the concave surface upwards to keep out dust, and to allow of the moisture falling back into the basin, so that the exposure to the sunlight may be prolonged), and allow it to evaporate in strong sunlight, testing it every few days by staining sections in it, until the nuclei and fibrous tissue are stained distinctly pink, whilst epithelial cells, · &c., are stained yellow. The best double staining is usually given before the fluid has evaporated down to half its bulk, and at this stage it is sometimes found that crystals of picric acid are deposited in the tissues. To obviate this, it is necessary to add 10 or 20 parts of distilled water. To prevent the growth of fungi, add from 2 or 6 drops of 1 in 20 carbolic acid solution to each ounce of the fluid; filter, and keep in a glass stoppered bottle."

Sections are stained in the following manner:—Alcohol must first be abstracted by allowing the specimens to float in water for some minutes. They are then removed one by one to glass slides and carefully spread out. A few drops of the staining fluid are added to each and allowed to stand for about five minutes. The excess of stain round the sections is then wiped off with a cloth, but the sections themselves are not dried. A little Farrants' solution is then added, cover-glasses applied and fixed by "ringing" with Dammar varnish or balsam.

If examined immediately the sections will not appear to be properly coloured, but this result will be satisfactorily completed in a few days, when the excess of staining fluid will be found to have become absorbed, and a most beautiful selective double staining will be brought about.

We now pass to a short consideration of the other carmine stains which are employed. These are given rather for the sake of completeness than for any advantage which they possess over lithium carmine. An exception must be made in the case of ammonium carmine, which is especially valuable for staining sections of the central nervous system.

Alum carmine—

Carmine. . . . . 1 part
Alum (5 per cent. solution) . . 100 parts.

The carmine is added to the alum, boiled for a quarter of an hour, and then filtered. A little carbolic acid should be added for preservation.

Sections are allowed to remain in the undiluted stain for about ten minutes. They are then washed in acidulated alcohol (see p. 76) until no more colour comes out, dehydrated, cleared and mounted in balsam.

Ammonium carmine—

Carmine . . . . 1 part
Strong ammonia . . . . . 1 part
Water . . . . 100 parts.

Place the carmine in a mortar and add sufficient water to make a paste; then pour in the ammonia, filter, and place a small piece of camphor in the bottle for preservation.

As already stated this stain is especially applicable
for sections of the central nervous system.

After removal from the hardening reagent (chromates)
the specimens should be dipped for about ten minutes
in a ·2 per cent. solution of chloride of palladium.

Sections must remain in the carmine for about ten
minutes, and be subsequently washed in water.

The medullary sheaths of the nerves are stained
yellow, the nuclei, ganglion cells, and axis-cylinders
being coloured a deep red tone.

Friedländer states that this dye is well adapted for
staining fresh bone specimens which have not been
artificially decalcified. It is thus very useful in specimens of rickets and osteomalacia.

Borax carmine (Grenacher).

Carmine	·5 parts
Borax	2 parts
Water	100 parts.

The mixture is placed in a porcelain dish and heated
to boiling, and a 5 per cent. solution of acetic acid is
added until the purple colour changes to red. The
solution is allowed to stand for twenty-four hours, filtered and preserved with a few drops of carbolic acid.

Sections are stained in a few minutes and are subsequently partially decolourized in acidulated alcohol.
The stain is useful for sections of the spinal cord and
brain. Its chief use is for staining specimens in
"bulk," that is, to stain preparations after they have
been hardened and before sections have been cut.

This method is generally used for portions of the
spinal cord and brain. They are allowed to remain in
the undiluted stain for from twelve to forty-eight hours,
according to the size of the specimens, and then from
three to twelve hours in the acidulated alcohol, until no

more colour can be extracted. They are next transferred to pure methylated spirit and absolute alcohol in succession, remaining in the last named not less than three hours, after which they are removed to about teñ times their bulk of chloroform or benzol until thoroughly saturated, and are finally embedded in paraffin in the usual way (see p. 52). Sections are cut, and after dissolving out the paraffin they are cleared and mounted in balsam.

Two other nuclear stains may be referred to here, namely, safranin and iodine green. They do not present any great advantages over hæmatoxylin and carmine, but may be used for the sake of variety.

Safranin is one of the best nuclear stains we possess, and is therefore especially useful in studying rapid cell-formation in morbid growths, or the process of karyokinesis. It is also an excellent stain for epithelial structures. It communicates a brilliant orange-red colour to nuclei. Its powers as a selective stain for colloid material will be again referred to in a future chapter.

In preparing the stain great care must be taken to procure a good sample; a reliable dye may be obtained from Dr. George Grübler, Baiersche Strasse 12, Leipsig.

The best formula is:—

Concentrated alcoholic solution of safranin   1 part
     ,,   aqueous   ,,   ,,   1 part.

Sections are allowed to remain in the stain for about a quarter of an hour, and are then washed in absolute alcohol until they become almost colourless, when they are removed to oil of turpentine for the purpose of clearing, and may then be mounted in balsam.

Iodine green.—This stain is best used in the form of a one per cent. aqueous solution. Sections placed in it stain almost instantaneously, and are improved by previous floating in water. After staining they are washed out in water, dehydrated in alcohol, and mounted in balsam.

With ordinary tissues it has peculiar differentiating powers producing three shades of colour in those elements which it selects. Thus epithelial cells and nuclei are stained dark blue; glandular structures dark green, whilst muscular fibres acquire a malachite green tinge. This dye has also a selective action for amyloid materia which it colours a rose-pink.

In order to produce a contrast in the parts of the tissue other than the nuclei, certain **general stains** are used with the above nuclear stains. Eosin and picric acid as contrasts to hæmatoxylin and carmine respectively have already been referred to. Eosin is also much used in the examination of blood, as will be explained in the chapter devoted to that subject.

A few other general stains will be mentioned here, which the reader may use at his pleasure.

Bismarck brown.—This dye, also known as **"vesuvin,"** is more used as a contrast stain in bacteriological work than with ordinary microscopic specimens. It is, however, very valuable for staining sections of bone and those exhibiting young granulation tissue. The most convenient form in which to use it is:—

Bismarck brown . . . . 2 parts
Alcohol . . . . 15 parts
Distilled water . . . . . 85 parts.

It yields a light oak tint which is very generally distributed over the mass of the tissue.

The sections should be allowed to remain in the stain for about a quarter of an hour, and then be swilled in water before dehydrating.

G

This dye forms a pleasing contrast stain to safranin or hæmatoxylin.

Rubin.—A 5 per cent. watery solution is the best strength to use. Sections are allowed to remain in the stain for about five minutes; are then washed in water and transferred to absolute alcohol; a good deal of the colour will now be taken out, and this should be allowed to continue until the specimens are light pink in colour; when they should be removed to oil of cloves and finally mounted in balsam.

Rubin colours the tissues a claret-red and thus presents a good contrast to hæmatoxylin; like eosin it picks out the fibrous tissue very beautifully. It is an excellent stain for the clubs of actinomyces (see p. 100).

Orseille.—The stain is made thus :—

    Absolute alcohol . . . . 20 parts
    Acetic acid . . .· . . . 5 parts
    Distilled water . . . . 40 parts.

Pure orseille is then added in sufficient quantity to produce a dark-red liquid.

The solution must be filtered before use. This dye is but little used except for staining the clubs of actinomyces, being introduced for this purpose by P. Israel. It is a fair contrast stain for hæmatoxylin, colouring the general tissue a deep crimson. Sections require to be left in the stain for about ten minutes, and are then washed in water, cleared and mounted in balsam.

Multiple staining.—The majority of sections which are now prepared in the laboratories are double-stained, as by this means the several parts of the tissue are demonstrated more clearly than when only a single stain is used. The two most common and most satis-factory combinations, namely, hæmatoxylin and eosin, and carmine and picric acid, have already been considered. Other arrangements of dyes may be employed according to the fancy of the worker, and the method of combination can easily be invented by referring to the dyes under the various heads of "nuclear" and "general" stains.

. In Lee's "Microtomist's Vade Mecum" the reader will find a large variety of combinations, the best amongst which are:—

Picro-carmine and iodine green.

Methylene blue and eosin.

Methyl green and eosin.

Methylene blue and Bismarck brown.

Fuchsine and methylene blue.

Treble staining.—The use of three colours is hardly of any advantage; the process becomes tedious and no fresh information can be derived from the addition of a third dye. To those who wish to make the experiment, we recommend Dr. Heneage Gibbes' process (slightly modified) as given in his work on "Practical Pathology and Morbid Histology," p. 51, as being the most satisfactory yet introduced.

The stains required are:—

Solution 1.—Lithium carmine (see p. 75).

Solution 2.—Rosanilin (Gibbes' formula, opus cit., p. 47).

"Place some of the crystals of rosanilin hydrochloride in a glass mortar and rub up with a little spirit, add more spirit until all the crystals are dissolved."

Solution 3.—Iodine green (see p. 80).

The method is carried out as follows:—

The sections are first stained in lithium carmine according to the process already described, and are then soaked in acidulated alcohol. A few drops of the solution of hydrochloride of rosanilin are then diluted with spirit, the sections immersed for two or three minutes, and subsequently removed to methylated spirit to wash off the excess of colouring matter. They are next placed in a dilute aqueous solution of iodine green.

Coming from spirit, they float on top of the watery solutions, and this in many cases, when the green stain is not required to be very deep, is quite sufficient. When a deeper stain is required, they must be immersed altogether, and allowed to remain for a minute or two, but it must be borne in mind that this colour cannot be washed out again if too deep, while the spirituous stain can, so that it is better to have a section apparently over-stained in the rosanilin solution, while it is even under-stained in the iodine green. After washing, the sections are dehydrated, cleared, and mounted in the usual manner. It will be found, however, that a good deal of the rosanilin will come out in the second immersion in spirit, and it is necessary to change it until no more colour comes away, otherwise the oil of cloves will become coloured, and from it the Canada balsam in which the specimen is mounted.

Dr. Gibbes adds:—" With the above-mentioned three colours, the most beautiful effect may be obtained, but it will take some time and practice to get the process exactly right, and proficiency in the matter can only be gained by experience. The results will be found to vary with the length of time the section is immersed in each of the two last colours and also with the strength of the solutions. The best results will also be obtained with material that has been hardened in chromic acid."

If sections of tongue are stained by this method, all the muscular fibres will be stained with the carmine, as also will the connective tissue, protoplasm of cells, &c.,

while all the nuclei in the superficial epithelium, serous glands, non-striped muscle tissue in the vessels and elsewhere are stained a brilliant green.

The most important fact demonstrated by this process is the different reactions shown by the various glands. In the mucous glands, although the epithelium lining the duct is stained in the same manner as the superficial epithelium of the organ, yet the secreting epithelium assumes a new colour differing entirely from either of those employed, namely, a sort of purple. In the serous glands, however, there is no such change, the protoplasm of the cells is stained more or less deeply with red, while the nuclei have taken on a green colour.

CHAPTER VI.

SELECTIVE STAINS.

UNDER the term "selective stains" we include those dyes which, for special reasons, pick out various components of the tissues (other than the nuclei) or bodies contained in them (micro-organisms). This selective power is exerted in some instances by a species of chemical reaction, such as the precipitation of reduced metal in the presence of light, as is the case with gold, silver, and osmium salts; or in another class of cases the singular production of colour by amyloid material. In other instances the selective action is due to the power of micro-organisms to retain dyes which are removed by suitable agents from the other parts of the tissue. Of this kind are the large and valuable series of aniline dyes. Since these have been introduced into microscopical technique an immense advance has been made in our knowledge of bacteria. If a section be placed in a solution of one of these dyes for about five minutes and then removed and placed in water, it will be found to be opaque, and coloured almost black, whilst at the same time a peculiar shimmering cloud is formed around it. But if the section be removed to alcohol, a large amount of colour is at once given up, the specimen gradually becomes paler, more transparent, and assumes a violet, blue, or reddish tinge (according to the dye used). When this stage is reached the process must be stopped and the section removed to cedar oil for clearing. The exact time at which this removal must be effected can only be learnt by experience.

Some of the aniline dyes, such as methyline blue, are frequently used as simple nuclear stains, but possess no advantage for this purpose over haematoxylin or carmine.

We will first deal with metallic stains. These are sometimes known as "impregnation methods," and are of two kinds. In a "negative impregnation" the intercellular substances alone are coloured, the cells themselves remaining colourless or very lightly tinted. In a "positive impregnation" the cells are stained whilst the intercellular spaces remain unaffected. This last variety, however, is very seldom used. The exact nature of the deposit formed in the intercellular spaces by the use of metallic stains is not known. Von Recklinghausen considered that the silver salt combined with a hypothetical intercellular cement substance forming a compound that blackened under the influence of light. Other authors are of opinion that the metallic salt combines with albuminous and saline liquids that surround the cells, and is then precipitated in simple intercellular spaces.

Nitrate of silver.—The most convenient strength for using this reagent is a half per cent. solution in distilled water. It is more frequently employed for physiological than for pathological work. It is especially useful for pathological conditions of the eye and for tumours of epithelial type. It can only be used when the specimens are perfectly fresh, and its employment is therefore limited to structures which have been removed during life. Very thin sections must be made which are washed in distilled water to remove chlorides, and are then placed in the reagent for about five or ten minutes. They soon become white, and as soon as a greyish appearance is noticed they must be washed in distilled water, and then removed to ordinary water and exposed to the daylight until they become brown. The sections may be preserved in spirit until wanted, or mounted at once in glycerine. Such specimens must be kept in the dark or they will become black and opaque.

Sections may be counter-stained with lithium carmine and picric acid, or other carmine dyes, care being taken, however, to avoid solutions containing free ammonia, which would dissolve out the silver. During impregnation by the silver salt, unless the sections or membrane be kept perfectly flat, the silver salt will be deposited unequally. To prevent this it is a convenient plan to pass the membrane over a porcelain ring, and fix it with another ring in the manner in which a skin is fitted on to a drum (Stirling).

A remarkable example of the positive impregnation by silver is exhibited by the condition known as argyria, in which silver salts are deposited during life in various tissues of the body, such as the Malpighian tufts of the kidney, in those patients who are taking silver by the mouth.

Chloride of gold.—In order to demonstrate the peripheral nerve endings, or the connective tissue corpuscles of the cornea, no reagent produces better results than chloride of gold. It is, however, extremely difficult to manipulate, and satisfactory specimens can only be obtained after long practice and with considerable exer cise of patience.

The same strength is used as for nitrate of silver, namely, half per cent. solution in distilled water; and the specimens must be perfectly fresh. A large number of methods for staining by chloride of gold have been introduced. Lee's "Microtomist's Vade Mecum" mentions no less than sixteen, but only the two yielding the best results will be described here, namely, Cohnheim's method, and Ranvier's lemon-juice method.

Cohnheim's method.—Small portions of the tissue immediately after removal from the body are placed in the chloride of gold solution of the above strength until they are thoroughly yellow. They are then removed to water acidulated with acetic acid, until a purplish tinge is assumed, by which time the gold is reduced. About four days are required for this process. Sections are cut and mounted in glycerine. Specimens thus prepared will only keep a short time.

Ranvier's lemon-juice method.—This is especially useful for demonstrating nerve endings. Portions of perfectly fresh tissue are soaked in lemon-juice, filtered through clean starchless muslin until they become transparent. This will require from five to ten minutes. They are then rapidly washed in water, and placed for about twenty minutes into one per cent. gold chloride solution. After which they are again swilled in water and then immersed in a bottle containing one per cent. of acetic acid in distilled water.

During this period they must be carefully kept in darkness. They are then exposed to the light for about twenty-four hours, after which sections may be cut by freezing and mounted in glycerine. Specimens thus prepared are not permanent. In order to procure such, after staining in the gold and washing in water, the specimens should be kept for twenty-four hours in a twenty per cent. solution of formic acid protected from light. The reduction of the metal is then complete, and after washing in distilled water, sections may be cut and mounted in glycerine.

Osmic acid.—The selective action of osmic acid on fat has already been referred to when speaking of this reagent as a hardening medium. It has also a selective property on medullated nerve fibres, which is made use of in the Pal-Exner method of staining the central nervous system. When used as a staining reagent it is kept as a one per cent. watery solution obtained by breaking the glass tube in which it is supplied and adding one hundred parts of distilled water. The solution is then transferred to a black bottle. Sections which have been previously soaked in some chromate solution for hardening purposes (as before stated alcohol must be avoided, when we wish to demonstrate fat) are immersed for about six hours in a solution of one-sixth per cent. of the acid, being meanwhile carefully protected from the light. They are afterwards washed in water and mounted in Farrants' solution. As a rule such specimens are at once hardened and stained in osmic acid (see p. 43), and after sections have been cut they may be counter-stained in picro-carmine.

We now come to consider the large subject of staining micro-organisms in animal tissues. During recent years no branch of microscopy has made greater advances than this; it has in fact almost become a science in itself, and the various processes introduced are so numerous as to need a volume of no mean size for a description of the whole of them.

Some of the methods are directed to the recognition of definite pathogenic organism, such as Ziehl's and Neelsen's for tubercle and leprosy bacilli ; other methods such as that of Gram, will stain the majority of micro-

organisms, whilst another, namely Kühne's, will colour
all known bacteria but two.

For a complete description of the various methods,
the reader is referred to the larger treatises on Bacteriology, but an effort will here be made to present in a
practical form such of the processes as are required for
staining the most important pathogenic micro-organisms.

Methylene blue.—As "Löffler's solution," this
dye has long been used for staining bacilli in sections,
especially those connected with enteric fever, diphtheria
and glanders. This method has now been in a great
measure superseded by that introduced by Kühne.
Löffler's is, however, much more. simple, and yields
very good results, and for general work is to be recommended, whilst for more extended study, and especially
in research Kühne's should be used.

Loffler's method.—

Löffler's solution has the following composition :—

Methylene blue (concentrated alcoholic
solution) . . . . } 30 parts

Solution of potash (1 to 10,000) . . 100 parts.

The best results are obtained from this solution, but
bacilli can be quite well demonstrated by a more simple
solution prepared thus :—

Methylene blue . . . . . . 2 parts
Alcohol . . . . 15 parts
Water . . . . 85 parts.

Sections are allowed to remain in either of the above
stains, undiluted, for at least twelve hours, a period of
twenty-four hours being better. A dilute solution of acetic acid is then made by adding about three drops of the
strong acid to a medium-sized capsule of water, or one
drop to a watch-glass full. In this mixture the sections
are washed after removal from the stain until they are

light blue in colour; they are then removed to absolute
alcohol when still more colour is removed, and afterwards they are cleared in cedar oil and mounted in
balsam.

It requires some practice to conduct the process of
washing out the blue or "differentiating" the specimens
properly; usually too much colour is removed both from
the tissues and the bacilli, or too much of the dye is left
in the tissues so that the bacilli can hardly be distinguished. Under such circumstances the margins of
cells or other constituents of the tissues especially
"mast-zellen" (see below), are very apt to be mistaken
for micro-organisms.

Kuhne's method.—There are several points in
which Kühne's method of staining micro-organisms with
methylene blue differ from those previously introduced,
notably in the addition of carbolic acid to the stain, and
the use of aniline oil for the purpose of dehydration instead of absolute alcohol.

In his general remarks on this stain, Kühne states
that he has succeeded in staining in the tissues almost
all forms of pathogenic bacteria, the only exceptions
being the bacilli of leprosy and mouse septicæmia.
The bacillus of glanders also stained in far fewer numbers than in cover-glass preparations. He suggests
that as leprosy bacilli will not colour by this method, it
might be made use of as a differential test in doubtful
cases of tuberculosis and leprosy.

For this method the following solutions are required:—
Solution 1.—" Carbol. methylene blue."
Absolute alcohol . . . . 10 parts
Methylene blue . . . . 1·5 parts.
Carbolic acid (5 per cent. aqueous sol.) 100 parts.
The dye is placed in a mortar and the alcohol poured

over it, the carbolic acid solution being gradually added during stirring.

Solution 2.—" Weak acidulated water."

Hydrochloric acid . .   . . 10 drops

Water . . . . 500 c.c.

Solution 3.—" Lithia water."

Water . . . . 10 c.c.

Concentrated aqueous solution of
lithium carbonate . . . . } 6 to 8 drops.

Solution 4.—"Aniline oil solution of methylene blue."

About as much methylene blue as will go upon the point of a pen-knife is rubbed up in a mortar with 10 c.c. of aniline oil. The whole of this mixture (the dye will not be wholly dissolved) is then poured into a small bottle. When required for use a few drops of the solution are added to a small capsule-full of pure aniline oil.

Solution 5.—" Aniline oil solution of safranin."

This is made in the same way as solution 4, using safranin instead of methylene blue.

Mode of procedure :—

Sections are removed directly from alcohol into the carbolic methylene blue (Solution 1), where they re-main for about half an hour. The specimens are next washed in water, and then differentiated by rinsing in weak acidulated alcohol (Solution 2), being afterwards dipped for a few seconds in lithia water (Solution 3) and again immersed in ordinary water. The process of de-colourisation must be carefully watched, and stopped when the specimens have assumed a pale blue tint. The time for this varies from a few seconds up to a quarter of an hour or more. Experience alone can guide one, and many disappointments will probably be experienced before success is attained. After remaining for about five minutes in water, the sections are dipped for a second or two into absolute alcohol, in which it is advisable to dissolve a pinch of methylene blue, and are then placed in methylene blue aniline oil (solution 4); owing to the alcohol they will spread out well in the oil, and dehydration will be complete in a few minutes. If the sections have become very pale.by this stage of the proceedings, rather more of the aniline methylene blue should be added than is directed above. After dehydration they are rinsed in pure aniline oil, which must then be extracted by thoroughly washing the specimens in xylol; it is very important that this be effectually done, otherwise the preparations soon become brown and almost worthless, two lots of xylol should be employed, the sections been transferred from one to the other. Finally they are placed on glass slides, and after draining off the superfluous fluid are mounted in balsam. The specimens may be counter-stained by immersing them in the safranin aniline oil (Solution 5) prepared in the same way as the aniline solution of methylene blue, the time required varies from two to ten minutes, according to the dilution of the stain and the material to be coloured. When removed from the stain they are immersed for a few seconds in pure aniline oil and then thoroughly washed in xylol.

Although Kühne states that all bacilli, with the exceptions named above, are coloured by this method, the author has attempted in vain to stain the bacilli of pseudo-tuberculosis.

"**Mast-zellen.**"—This seems to be the most suitable place to refer to these bodies. In sections of organs affected with diphtheria, glanders, and other processes, which have been stained by methylene blue or other aniline dyes, small round or spindle-shaped bodies are often met with ; they are about twice the size of lymph cells, and consist of coarsely granular protoplasm with a well-marked nucleus. The granules stain with the aniline dyes, but the nucleus remains uncoloured, appearing as a bright spot in the midst of the deeply-stained protoplasm. These bodies occur in large numbers in the connective tissue, especially in the mucous membrane, and are generally situated in the neighbourhood of the vessels. Their physiological and pathological significations are not known, they are very numerous around rapidly growing tumours and also in elephantiasis.

" Mast-zellen " have hitherto been chiefly described by foreign observers, and in England they are generally known under the term "plasma cells." They are very liable to be mistaken for colonies of micrococci by those unfamiliar with them, but may be distinguished from the latter by the presence of the nucleus and by the irregularity in size of the component granules.

"Gram's method of staining."—This process is the one most generally adopted for staining micro-organisms, it will stain all the most important pathogenic bacteria, with the exception of those associated with cholera, typhoid fever, and glanders.

The following solutions are required :—

Solution 1.—Methyl aniline violet. This should always be freshly prepared. A few drops of pure aniline oil are placed in a medium-sized test-tube, which is then nearly filled with water and thoroughly shaken, the thumb being placed over the mouth of the tube. The emulsion is filtered through a double-folded paper, and poured through a second time if the filtrate be not quite clear. A little of this aniline water is placed in a capsule, and to it is added, drop by drop, a concentrated alcoholic solution of methyl violet, until the mixture just becomes opaque. This stage is most easily recognised by placing the capsule on the edge of a piece of filter paper, and adding the dye until the margin is almost, but not quite, obscured.

Solution 2.—Gram's solution.

Iodine . . . .   1 part
Iodide of potassium   . . . .   2 parts
Water . . . .   300 parts.

The sections are transferred from dilute spirit to methyl-aniline violet, where they are allowed to remain for five or ten minutes; after staining they are transferred to Gram's solution until they become brown and opaque; one minute is usually long enough. They are next removed to 60 per cent. alcohol, and gently moved about by means of a needle until no more colour can be removed. Professor Crookshank's procedure may also be adopted which is to transfer the specimens two or three times backwards and forwards between oil of cloves and the alcohol. When completely decolourised, the sections may be counter-stained by placing them in absolute alcohol to which two or three drops of a saturated alcoholic solution of eosin have been added; they are at the same time dehydrated, and after clearing in cedar oil may be mounted in balsam.

One caution is here necessary: if the specimens have been embedded in celloidin previous to cutting, oil of cloves must not be used for decolourising, and it is better to dehydrate in 95 per cent. rather than in absolute alcohol.

Weigert's modification of Gram's method.— Weigert and Kühne were the first to observe that after staining bacteria in methyl violet and treating with Gram's solution and decolourising with alcohol, much of the colour was removed from the micro-organisms as well as from the tissues. These observers then proposed to use aniline oil for decolourising, and Weigert consequently introduced the following modification of the method just described.

The sections are first stained in lithium carmine as described on page 75, decolourised with acidulated alcohol and dehydrated with absolute alcohol. The sections are then transferred one by one to glass slides, and each is treated in the following way :—

A couple of drops of methyl aniline violet freshly prepared (see above) are placed on the section and allowed to remain for about five minutes, the excess is poured off, and a small quantity of Gram's solution poured on; after the lapse of one minute this is likewise poured off and all moisture removed by gently pressing a folded filter paper on the specimen. Pure aniline oil is now added, and made to flow gently too and fro by tilting the slide; the colour will be removed at first rapidly and then more slowly, and fresh portions of oil must be used until no more colour comes away. It occasionally happens that the section adheres to the glass all round, but not in the centre, so that some of the stain may be thus retained, if such be the case, a corner of the section must be gently raised with a needle so as to permit some of the oil to flow beneath it, care being taken to prevent the whole section from being floated away.

Decolourisation being complete, two or three lots of xylol are poured over the section, a light cloud will be observed at first, and the xylol must be used until this has quite disappeared; unless this part of the process be thoroughly carried out, the section will speedily turn yellow. A few seconds are allowed to elapse for the xylol to evaporate, and finally a drop of balsam is placed on the specimen and a cover-glass applied.

The lithium carmine solution is very apt to develop micro-organisms, and errors often arise on this account, the observer imagining that he has an extremely fine preparation. Some sections, such as of cirrhosis of the liver, known to be free from germs should therefore be passed through the above process so as to ascertain if the carmine solution be untainted.

The above process is particularly suitable for staining anthrax bacilli and sections of ulcerative endocarditis. In the latter not only are the masses of micrococci beautifully demonstrated, but a delicate purple tint is left in the layers of fibrin.

Neelsen-Ziehl method for staining tubercle bacilli.—In order that this process may be carried out as successfully as possible, the specimens should be hardened in alcohol, or if any of the chrome solutions have been used, the hardening material must be thoroughly extracted by water and the specimens soaked in alcohol for two or three days. Sections which have been cut in celloidin or paraffin will also be found more satisfactory to manipulate than those which have been frozen, as these last are very apt to curl up in the acid, and it is difficult to spread them out again.

As stated in the Chapter on the " Examination of Sputum," this method depends upon the property which the tubercle bacilli possess of resisting for some time the action of dilute acids, which distinguishes them from all other bacilli, except those of leprosy which act in the same manner ; but as in this country leprosy is fortunately a very rare disease, this method of staining may be said to be an absolutely distinctive one. Tubercle bacilli also stain by Gram's and Kühne's methods, but these are never used for the purpose. I have used the double name for this process, for although introduced independently by the two observers, the processes are essentially the same.

The solutions required are :—

Solution 1.—Carbol. fuchsine (Neelsen-Ziehl solution).

Fuchsine . . . . . . 1 part

dissolved in a 5 per cent. watery solution of

Carbolic acid . . . . . 100 parts

Alcohol . . . . 10 parts.

Solution 2.—Sulphuric acid, 25 per cent. watery solution.

Solution 3.—Methylene blue :—

Methylene blue . . . . 2 parts

Alcohol . . . . 15 parts

Water. . . . . 85 parts.

The sections having been soaked in dilute alcohol, are placed in the fuchsine solution contained in a covered capsule, where they should remain for at least an hour, or preferably for twelve hours. They are then decolourized in the sulphuric acid. This process is carried out thus :—Sections are allowed to remain in the acid for about thirty seconds, and are next washed in a large capsule full of 60 per cent. alcohol, being gently moved about with a glass rod. Part of the colour will be probably restored. If so, they must be returned to the acid for about ten to fifteen seconds, then again washed in the alcohol, this process being repeated until only a faint lilac tint remains. They are next counter-stained in the methylene blue solution. An immersion of half a minute will be sufficient, after which they must be again washed in water. The blue stain is rapidly removed by the water, and in about half a minute the

H 2

sections should be transferred to absolute alcohol. Still more colour will then be abstracted, and the specimens must therefore be watched, or they will become completely decolourised. As soon as they assume a light blue tint they should be removed to cedar oil for clearing, and finally mounted in balsam. The bacilli will be found to be stained a brilliant red, whilst the nuclei of the tissues are coloured blue. This stain is very permanent lasting for years.

A more simple but not so reliable a method consists in immersing the sections for several hours in **Gibbes' double stain,** and subsequently washing out the excess of colour in methylated spirit, dehydrating in absolute alcohol, clearing in cedar oil, and mounting in balsam. Gibbes' double stain may be procured at any of the microscope makers. Its composition is rather complicated :—

Rosaniline hydrochlorate . . . . 2 parts
Methylene blue . . . . 1 part.

The mixture must be triturated, and 3 parts of aniline oil mixed with 15 parts of rectified spirit slowly added, followed by 15 parts of distilled water.

Methods of staining actinomyces.—This fungus has been more frequently met with during the last few years than it was formerly, not because cases have been any more numerous, but on account of discharges from the chest and other parts being more carefully examined in doubtful cases.

For demonstrating the fungus in sections I have found the following method most satisfactory. The sections are first stained in rubin (see p. 82) for ten minutes, are then washed in water and placed in absolute alcohol for about three minutes. If the sections are seen to become very pale they must be taken out sooner. They are next transferred one by one to glass slides, and stained according to Weigert's modification of the Gram method (see p. 96). On examination the clubs of the actinomyces will be found to be stained red, whilst the mycelium appears violet.

Plauts' method, which is the one usually adopted, is as follows :—The sections are placed for two hours in Neelsen's solution (see p. 99). The time of staining may be shortened to twenty minutes if the fluid is maintained at a temperature of 40° C. in an incubator. The sections are next washed in water and immersed for ten minutes in a saturated alcoholic solution of picric acid. After washing for twenty minutes in water they are transferred for about fifteen minutes to 50 per cent. alcohol. In about a quarter of an hour they will be properly decolourised, and may then be dehydrated in absolute alcohol, cleared in cedar oil and mounted in balsam. By this method the clubs only are stained, the mycelium remaining uncoloured.

I now come to consider a small class of selective stains which are dependent upon micro-chemical reactions. These are practically confined to two, namely, the methods for demonstrating waxy degeneration (amyloid reaction), and for the exhibition of iron in the tissues.

The amyloid reaction.—There are several methods for showing this reaction, by means of which portions of tissue which have undergone waxy degeneration are clearly distinguished by change of colour from the normal portions. The chief processes are as follows :—

1. A thin section of the tissue is placed on a slide, and a drop of an aqueous solution of iodine upon it.

The specimen may then be mounted in the iodine-mounting fluid (see p. 109). The edge of the cover-glass being fixed by a ring of Dammar varnish. On examination the affected portions appear as dark yellow or brown, the remainder of the section being light yellow.

2. This was Virchow's original method. The sections as before are dipped into a watery solution of iodine, and are then conveyed to a 2 per cent. solution of sulphuric acid. As soon as the blue colourisation appears, the sections are swilled in water, and mounted in Farrants' solution.

3. *Methyl aniline violet.* The stain is made as described on p. 95. The sections are placed in it for about five minutes, are then washed in water for twenty-four hours, and mounted in Farrants' solution. The amyloid substance will be found to be coloured red, the rest of the tissue being a dull blue. The preparations will not keep well in the mounting medium unless they have been previously fixed for half an hour in a one per cent. solution of perchloride of mercury.

. 4. *Iodine green.* The sections are soaked in this stain (see p. 80) for ten minutes, then washed in water and mounted in Farrants' solution. Very beautiful results are thus obtained, the affected portions being coloured a rose pink, whilst the remainder assume a bluish green colour.

Iron reaction.—Iron is occasionally found free in the liver and other organs in diseases in which excessive destruction of red blood corpuscles is brought about. Dr. Mott in 1889, before the Pathological Society, showed some beautiful sections of liver from a case of pernicious anæmia. The iron was seen in the form of blue granules (converted by ferrocyanide of potassium into Prussian blue) in the cells and capillaries of the portal zone.

The method for obtaining such specimens is as follows:—The organs must be hardened in alcohol, and after sections have been cut, they are placed in a solution of ferrocyanide of potassium acidulated with hydrochloric acid. In a very few minutes the sections will turn blue, and may then be removed and mounted in glycerine. An objection to this method lies in the partial solubility of the prussian blue, which escapes and tinges other parts of the tissue a similar colour.

More exact definition of the deposit may be obtained by exposing the sections to a watery solution of sulphide of ammonium, which precipitates the iron as dark greenish granules, or by placing them in a solution of sulphocyanic acid, when the granules will be of a blood-red colour.

Before concluding the chapter, this seems to be the most suitable opportunity for referring briefly to the principle method which is employed for demonstrating the **karyokinetic figures in nuclei,** it is known as Flemming's method.

The organ to be investigated must be removed from the body immediately after death ; the larynx, after an operation for excision for instance, is very suitable for this purpose. The tissues must then be "fixed," so that the various components are made to retain their living structural appearance. Flemming's fixing mixture has the following formula :—

Chromic acid (1 per cent. solution) . . 15 parts
Osmic acid   (2 „   „   „   ) . . 4 „
Acetic acid . . . . 1 part.

Small pieces of the tissue are placed in this solution for two days, and are then washed in a stream of running water for at least an hour. They are afterwards hardened in alcohol, for which purpose three days are required. They may then be embedded in celloidin or paraffin (the former being preferable) and sections cut. The best staining fluid is undoubtedly safranin (see p. 80), after which they are washed in acidulated alcohol (see p. 76), then dehydrated in absolute alcohol, cleared in oil of cloves and mounted in balsam.

Most of the nuclei will by this method be almost decolourised, whilst the karyokinetic figures will be stained a brilliant red.

Another process for demonstrating these figures which produces very pretty results, but not quite so well defined as with Flemming's method, is after hardening the tissues in alcohol to stain them by Gram's method (see p. 95), when those nuclei which are undergoing division will be coloured a deep purple, whilst the others will be almost decolourised.

CHAPTER VII.

Clearing and Mounting.

THE final stage in the preparation of microscopic specimens is perhaps the most difficult, and requires more
practice than any of those previously described.   It
not unfrequently happens that after a section has been
successfully cut and stained, it is irretrievably damaged
in spreading it out on the slide.   Skill in manipulation
can only be obtained after much practice, and each
worker will probably arrange methods of his own, but
a few general hints may be of service in indicating the
more common errors and modes of correcting them.

   In order that a section may be satisfactorily examined
its interstices must be filled by some material having a
refractive index higher than that of air, in order to
prevent the irregular reflection resulting from the want
of homogeneity.   Before these reagents can permeate
the tissues, the sections have to be cleared.   The liquids
employed for this purpose have also a high index of refraction, by which the tissues are rendered transparent,
and at the same time the alcohol which has been used
for dehydration is removed, and the balsam or other
resinous media can easily flow in.

   Two varieties of mounting media are employed,
namely, fluid and solid, and the methods for employing
these differ very much from one another.   We will consider first the use of **fluid media.**

   After the completion of the staining process the

sections are placed in water; each one is then removed
to a glass slide by means of the section lifter. The
specimen must be carefully spread out on the lifter by
means of needles, and then very cautiously pulled up
out of the water. Unless this is slowly done the section
is nearly sure to curl up, or float off from the lifter.
Having successfully removed the latter, with the former
adhering to it, the lifter is laid on the centre of the
glass slide, and one corner of the section having been
fixed by means of a needle, the lifter is cautiously
pulled from under it and not the section from off the
lifter.

Sometimes in spite of all care, especially if the section
had been cut by freezing, the specimen will obstinately
refuse to lie properly on the lifter, whilst this instrument
is being removed from the water. A very good expedient is to place a clean slide in the dish containing
the water, and then to spead a section on it by means
of a needle, by gently steadying it thus, the slide may
be raised, and so the specimen lifted out lying perfectly
flat.

The slide is now tilted, so as to allow superfluous
fluid to strain off, and as much water as possible is
removed by placing a piece of folded filter paper near
the section, or if the object be not too delicate the paper
may be gently pressed on it. A drop of the mounting
fluid is then added, and a cover glass laid on it in the
following manner. The cover glass is taken between
the finger and thumb of the left hand and placed with
one edge downwards, just beyond the drop of fluid.
The upper edge of the glass is supported by a needle
which is gradually lowered until the glass comes in
contact with the liquid. If this be carefully done no
air bubbles will be included, but if this should happen

the glass must be lifted up, and a fresh one applied. The amount of fluid should be sufficient so as just to occupy the space beneath the cover-glass, as any excess interferes with the subsequent application of the fixing material. This quantity can only be gauged by practice.

The cover-glass having been placed in position, it must be fixed by one of the cementing materials presently to be mentioned. For this purpose the glass slide should be held in the left hand. Any excess of mounting fluid which has escaped must be removed with a soft cloth. A glass rod is then dipped in the cementing fluid and gently passed round the edge of the cover-glass. This should be done by holding the rod a fraction of an inch above the glass so as to avoid touching it, but at the same time be near enough to guide the fluid. The rod will probably have to be dipped in the liquid two or three times before the circuit of the glass is complete. It is needless to add that only a very narrow portion of the cover-glass must be included in the ring; but the same precaution need not be taken with the glass slide.. As the .cement hardens a second or third coating should be applied, so as to form a raised border.

The most common fluid media employed must now be considered.

Glycerine.—This is an excellent medium for sections which have been previously hardened and are stained in dyes which are easily extracted by means of alcohol. It is also suited for specimens such as those containing fat which would be rendered useless by dehydration, and further, for those sections which have been treated with metallic salts.

It has the great disadvantage of causing fibrous tissue to swell up, and so lose its characteristic structure.

In mounting fresh specimens this difficulty may be overcome by covering them the first day with equal parts
of glycerine and water, and on the following day removing the cover-glass and adding pure glycerine.

This reagent is much used for preserving parasites.
For this purpose such specimens should be immersed in
equal parts of glycerine and water, contained in a small
covered capsule for twenty-four hours. On the second
day a small portion of the liquid must be removed with
a pipette, and a few drops of glycerine added. This
process being repeated day by day until the parasite is
surrounded by the pure reagent.

Another excellent way of treating delicate objects is
to employ a mixture of glycerine, alcohol and water.
By the evaporation of the alcohol the liquid gradually
increases in density, and after some time a cover-glass
may be applied and fixed, or the object brought into
pure glycerine.

Lee recommends the following mixture :—

Glycerine . . . . 1 part
Alcohol . . . . 1 ,,
Water . . . . 2 parts.

Glycerine jelly.—This should be procured already
prepared. The bottle containing the reagent is placed
in hot water until the contents are fluid. The object
having been placed on the slide and dried, the latter is
slightly warmed and a drop of the jelly added. The
cover-glass is then applied and gently pressed down,
any air bubbles being expelled by pressure with a glass
rod.

The jelly soon solidifies and fixes the glass, but it is
more secure to protect the preparation with a ring of
cement. This reagent is useful for vegetable substances
and for parasites.

Glycerine mounting fluid :—

Camphor water . . . . . . 2 parts
Glycerine . . . . 1 part
Pure gum Arabic . . . . . 11½ parts.

This preparation is chiefly used for sections which have been injected with Prussian blue. The method of using it does not differ from those of the other fluid media.

Farrants' solution.—Farrants' solution being very difficult to prepare is best procured ready made. It is an extremely useful reagent combining most of the advantages of glycerine, with but few of its disadvantages. Sections preserved in it, however, are apt to become cloudy after some years.

Sections are mounted in the usual way, and after some days the fluid dries at the edge of the cover-glass, so fixing it ; but it is better to add a ring of Hollis' glue, or zinc-white cement.

It is especially useful for mounting fresh specimens or sections which have been stained to demonstrate the amyloid reaction (see p. 101).

Iodine mounting fluid :—

Glycerine . . . . 6 parts.
Liquor iodi (B.P.) . . . . . 3½ „
Water . . . . 6 „

The ingredients are mixed and six parts of gum Arabic added.

This preparation is only employed for sections stained in iodine. The method of using it is the same as for glycerine.

Two other fluid media may be mentioned, namely, **acetate of potassium** in a saturated solution for mounting vegetable tissues or those exhibiting fatty change; and **castor oil** for mounting crystals which are soluble in Canada balsam.

CEMENTING MATERIALS.

A section having been mounted in one of the fluids just mentioned, a ring of cement is applied to fix the cover-glass. If a square one has been used the cement is applied in the manner already described (see p. 107); but if the glasses are circular a Shadbolt's turn-table may be employed.

The most usual materials employed are Canada balsam which must be liquid enough to run easily, or Dammar varnish (see p. 115).

Other cementing substances are:—Brunswick black, Hollis's marine glue and gold size. Any of the above may be supplemented by a ring of zinc white.

All these materials are best secured ready prepared, as their composition is complicated and not readily carried out.

For sections mounted in glycerine, Dr. Woodhead ("Pathological Histology," p. 80) strongly recommends the method suggested by Dr. Marsh, namely, a solution of gelatine; the rationale of the process being that gelatine readily mixes with the glycerine in its immediate neighbourhood. The solution is prepared by placing a small quantity of gelatine in a narrow glass beaker, covering it with water, and allowing the gelatine to take up as much of the water as it will. Any superfluous water is poured off and the mixture is heated and three or four drops of creosote are added to each ounce of the fluid. The mixture will then solidify, and each time it is needed, the bottle containing it is immersed in a cup of warm water to render the contents fluid. All superfluous glycerine must be removed by the aid of a camel-hair brush and a damp cloth. A ring of the gelatine fluid is painted round the edge of the cover-glass; as soon as this is set it is painted over with a solution of bichromate of potash, made by dissolving ten grains of that salt in one ounce of distilled water. Dr. Marsh recommends that "this application of bichromate of potash, should be made in the daytime, as the action of daylight upon it, in conjunction with the gelatine, is to render the latter insoluble in water."

The preparation is finished by washing it over with methylated spirit to remove all the glycerine, and then a ring of zinc-white is applied.

Mounting in solidifying media.—This is accomplished by preserving the sections in a transparent refracting material which solidifies on drying into a vitreous mass. Various resins are employed, dissolved in a fluid which easily evaporates.

As these substances are insoluble in water this must be removed. This process is known as "dehydration."

After staining and washing, the sections are placed in absolute alcohol, by which means all the water is rapidly extracted. The time usually required is five minutes, but if the specimens are thick a longer period may be necessary.

Great care is needed in the dehydration of sections embedded in celloidin. As this material is soluble in absolute alcohol it is advisable to employ 95 per cent. alcohol to abstract the water, allowing about ten minutes for the process.

Very delicate specimens may be treated thus :—A clean cover-glass is placed at the bottom of a shallow dish containing water, and the sections floated on to it. By means of needles one specimen is spread out on the cover-glass, which is removed with a pair of forceps, and two or three lots of absolute alcohol poured over it. The spirit is allowed to evaporate, and a clearing solution applied in the same way. Finally the section is mounted in balsam.

In the ordinary course of events each specimen is removed from the alcohol on a section lifter and immersed in one of the clearing agents. Owing to the rapid escape of the alcohol the section flattens out and floats on the surface of the fluid, but it should be gently forced below the surface with a needle. The specimen will soon become perfectly transparent. If any cloud appears, or if, white spots are seen (this being best tested by viewing it against a black surface, such as the sleeve of a coat), dehydration has been inefficiently performed, and the section must be again placed in the absolute alcohol.

When perfectly transparent each section is removed to a clean glass slide by means of a lifter which is laid flat on the slide. One corner of the specimen is then fixed with a needle, and the lifter cautiously withdrawn. If this has been done with skill the section will be almost flat. If any part of it has become rolled up, or wrinkled, this defect must be remedied by means of two clean needles, the one being used to fix the section, and the other to remove the creases. This is rendered easier by having an excess of the clearing reagent on the slide.

The slide is now tilted to allow the clearing fluid to run off. This must be done gently, and if necessary the section steadied by a needle to prevent its slipping off. After draining, the superfluous oil is removed with a cloth or a folded filter paper may be laid on the section, and gently pressed.

A drop of Canada balsam or Dammar varnish is then

placed on the section and a clean cover-glass applied in the same manner as described on p. 106. By this means the intrusion of air bubbles is prevented as far as possible, but should any occur an attempt may be made to remove them by gentle pressure on the cover-glass. Very often this is unavailing, and it is better to leave them alone, as they are generally expelled by the contraction of the balsam, and undue pressure may either damage the section, or break the cover-glass.

Occasionally after two or three days gaps are noticed beneath the cover-glass which may even trespass upon the section. The balsam must then be liquefied by gently warming the slide, and a drop of balsam placed on the edge of the cover-glass nearest the gap. A little pressure will drive out the air, the resin taking its place.

For some days after mounting, the slides should be allowed to lie flat, protected from dust, and when the balsam has thoroughly set they may be stored edgeways. Slides should be labelled as soon as finished, the description including the name of the organ, or nature of the tissue, the disease, method of staining and date.

We now have to consider the various clearing reagents in use.

Any of the essential oils will answer the purpose, but those most commonly employed are, oil of cloves, creosote, cedar wood oil, xylol, and oil of bergamot.

Oil of cloves.—Although more generally employed than any of the others it has several disadvantages; it certainly produces excellent results, but has a clinging and penetrating odour, rapidly dissolves celloidin, and removes the aniline dyes, picric acid, and eosin. If sections be exposed too long to its action, they become

I

rigid and brittle, and should therefore be removed directly they are transparent.

Creosote.—I prefer this to any of the others for ordinary microscopic work, except for use with specimens stained in the aniline dyes. Its odour is not disagreeable, sections may remain in it for any length of time without becoming brittle, and it does not dissolve celloidin.

Cedar wood oil.—For bacteriological purposes this is undoubtedly the best clearing reagent. It does not dissolve celloidin, or abstract the aniline dyes. It has, however, rather a tendency to stiffen the sections, especially if they are left in it for some time.

Xylol.—This is not very often used, except in special methods, such as Weigert's modification of the Gram method, and Kühne's processes for the demonstration of bacteria in the tissues.

Oil of bergamot.—This is a good clearing agent, and has the advantage of not dissolving celloidin or affecting the aniline dyes. It has, however, a very disagreeable smell.

Turpentine, oil of cajeput, and **oil of origanum,** are sometimes employed, but possess no advantage over the others.

Practically the only two resins employed for mounting specimens are Canada balsam and Dammar varnish.

Canada balsam.—This is suitable for all specimens which will bear dehydration in alcohol, and clearing with an essential oil. It may be procured dissolved in various fluids. The most common being turpentine or chloroform, or a mixture of the two. Benzol also is sometimes employed.

Any of these solutions are suitable for ordinary sections, but for bacteriological work, or for sections stained in the aniline dyes, xylol is decidedly the best solvent. Balsam dissolved in xylol may be procured from most of the firms that supply microscopical apparatus, or any of the ordinary solutions may be evaporated to dryness over a water bath, taking care that the balsam does not get over-heated and become brown. The dried resin is dissolved in xylol. The consistency of the solution should be such that it flows easily, that is to say, it should be nearly as thin as glycerine.

Taking everything into consideration this is the best kind of balsam to use. It should be kept in a special "balsam bottle," which consists of a wide-mouthed bottle furnished with a ground glass cap and a small glass rod lying in it.

Dammar varnish.—It is also well to procure this ready made, but its composition is as follows :—

Gum Dammar . . . . . 2 parts
Gum mastic . . . . . . 1 part
Turpentine . . . . . . 4 parts.
Chloroform . . . . . . 2 „

The gums are mixed with the solvents until they are dissolved, and the mixture filtered through cotton-wool. This preparation has the advantage of being quite colourless, and is therefore useful for photo-micrography. Sections preserved in it do not keep well, becoming after a time cloudy and granular. It must not be used for sections stained in the aniline dyes, or for those containing micro-organisms.

Breaking down old specimens.—In the course of preparation of a number of specimens a certain proportion will from one reason or another be spoilt. The glass slides and cover-glass need not be wasted, but a jar should be kept into which waste alcohol should be thrown, such as that used for dehydration. Into this rejected slides may be put from time to time, and after soaking for a week or two may be removed, and will then be quite suitable for another occasion.

At other times valuable specimens may be damaged in some way; the cover-glass may be broken, or the sections may have faded. A section which has been mounted in glycerine is easily released by cutting through the cementing material with a knife, and then placing the slide in water, when the section will float off. It may then be stained, and mounted again in the usual way.

One which has been preserved in balsam must be immersed for some time in chloroform. After some hours the section will be freed, and should then be allowed to lie in absolute alcohol for a day. It may subsequently be restained and remounted.

CHAPTER VIII.

COMPLETE PROCESSES FOR THE PREPARATION OF SECTIONS.

IN this chapter it is proposed to give as briefly as possible two methods of preparing sections. In the preceding chapters all the processes have been considered in detail, and should be carefully studied before this chapter is consulted. Whilst working I have found it convenient to have a summary of the various stages through which the specimens have to pass before they are complete, and give two of them here; others may be compiled by referring to what has already been described.

Method I.—Harden in Müller's fluid, cut by freezing, stain in hæmatoxylin and eosin, mount in Canada balsam.

1. Place the specimen removed as soon after death as possible, in a bottle containing several times its bulk of Müller's fluid. Change in twenty-four hours, merely pouring the fluid off without removing the specimen from the bottle. Change again on the third and seventh days, and at the end of each week until the sixth. Examine to see if the specimen is sufficiently hardened, if this be the case, transfer to dilute spirit (fifty per cent.) until wanted, if not, put back in Müller's for another week.

2. Place the specimen in water until it sinks, and the washings are no longer coloured. Transfer to solution

of dextrin (see p. 48) for twenty-four hours.   Remove the specimen by means of forceps to the plate of an ether freezing microtome.   Work the spray cautiously so as not to overharden.   Float the specimens off the knife into water.

3. Take a dilute solution of hæmatoxylin and allow sections to remain in it for several hours, or in a strong solution for about three minutes (p. 73).   Wash the sections in water and allow them to float exposed to the light for two or three hours.

4. Transfer to absolute alcohol, containing a few drops of alcoholic solution of eosin, and allow the section to remain for five minutes; or soak it in a watery solution of eosin for ten minutes and then place in absolute alcohol for three minutes.

5. Clear in creasote.

6. Mount in Canada balsam.

Method II.—Harden in alcohol, embed in celloidin, stain in carmine and picric acid, mount in Canada balsam.

1. Cut small portions of tissue and place them in absolute alcohol.   Change this solution in twenty-four hours.   The specimens will be hardened in from two to four days, depending upon their size, and may be left in the alcohol for an indefinite period.

2. Place the specimens in equal parts of absolute alcohol and ether for twelve hours; transfer to celloidin of the consistence of mucilage for twelve hours; place each specimen on a dry cork, and allow celloidin to set; then immerse in a large jar containing eighty per cent. alcohol, for at least twelve hours.   Select one specimen and place in the clamp of the microtome.   Be careful to have the knife and specimen constantly covered with an excess of dilute alcohol (fifty per cent.).   Transfer

sections when cut by means of a camel's hair brush to water, where they will spread out, then keep them in dilute spirit.

3. Place the sections in lithium carmine (p. 75) for ten minutes. Wash in acidulated alcohol (p. 76) until the sections are a light pink, or until no more colour comes out.

4. Transfer the sections to absolute alcohol containing a few drops of a saturated alcoholic solution of picric acid for five minutes.

5. Clear in cedar oil. If after the preparation of one specimen the contrast-stain is found not to be strong enough, more picric acid must be added to the alcohol. If white patches are noticed in the sections, dehydration has not been properly accomplished, and the sections must be put back into the absolute alcohol.

6. Mount in Canada balsam.

CHAPTER IX.

METHODS OF PREPARING SECTIONS OF THE CENTRAL
NERVOUS SYSTEM.

IN order to prepare sections of the brain or spinal cord,
very complicated and rather difficult methods have to
be adopted.   It has, therefore, been thought advisable to
devote a separate chapter to their consideration, and to
describe in detail the various processes from the time
that the cord or brain are removed from the body, until
the sections are complete.   As the examination of the
spinal cord is more frequently required than that of the
brain, the following description will be understood to
apply to the cord unless otherwise stated :—

Hardening.—The spinal cord should be removed
from the body as soon after death as possible, and there
will be less chance of injury if it be removed from the
front instead of from the back as is usual.   The method
of thus doing not being generally known, a brief outline
of the procedure is here given.   It is as follows :—The
viscera having all been removed, the inter-vertebral substance above and below the second lumbar vertebra is cut
through, the pedicles of the vertebra divided with bone-
forceps, and the body then wrenched out with lion forceps.
A specially devised chisel having a point projecting from
its cutting edge is inserted with this point in the verte-
bral canal, and is driven with a mallet through the
entire line of pedicles on the right side to the base of the
skull; a like manœuvre is performed with a similar

chisel on the left side. The bodies of the vertebræ having been removed, the theca is fully exposed and is slit up with probe-pointed scissors. The cord is removed by dividing the nerves on either side, and those of the cauda equina and gently lifting it with the membranes out of the canal.

Transverse cuts are made at intervals of half an inch if the whole cord is intended for preservation or the diseased portion is separated from the healthy.

The best fluid for hardening purposes is a solution of bichromate of ammonium (two per cent.) or Müller's fluid. The former is more rapid in its action, but the latter more regular, and is therefore to be preferred. The cord should be suspended in a tall jar, which should be large enough to contain sufficient fluid to measure several times the bulk of the specimen. The hardening reagent must be changed at the end of twenty-four hours, on the third and the seventh days, and at the end of each week until the process is completed, the jar in the meanwhile with its contents being kept in a cool place and in the dark. Five or six weeks will probably be required, and the cord should be tested from time to time to see if it is properly hardened. To do this, a small section of the specimen should be taken, and placed in water, it ought to remain flat, if it curls up it is not yet ready.

Exposure to Müller's fluid generally renders the tissue brown in colour. When sufficiently hardened, the cord is washed in water and steeped in eight per cent. alcohol, where it may remain until arrangements are made for cutting. For the sake of convenience the specimen may be cut into segments, which can be placed in small bottles carefully labelled.

Cutting.—Sections may be cut by freezing, but this

will be found by no means satisfactory, and it is far better to embed the cord in celloidin or paraffin. The choice of these two will depend upon many circumstances, in a great measure upon the microtome it is intended to employ. If a series of sections are to be cut by the Cambridge Rocking microtome, paraffin is undoubtedly to be preferred, but if any other microtome be selected celloidin should be chosen.

The method of embedding in paraffin does not differ from that described on p. 52. For celloidin, after removal from the hardening reagent the specimens are placed in absolute alcohol for two days, and are then transferred to equal parts of ether and alcohol, for at least twelve hours. They are next immersed in a thin solution of celloidin for some hours, and afterwards in another solution about the consistence of mucilage. After another twelve hours they are deposited upon pieces of cork a little larger than the specimens, and when the celloidin has commenced to set, the corks are thrown into a large jar containing eighty per cent. alcohol, care being taken that the celloidin is below the surface of the liquid. In twelve hours, sections may be cut by Richert's or Schanze's microtome.

Hamilton has introduced a method by which sections, which have been embedded in celloidin, may be cut on a freezing microtome. The specimens must be removed from the corks and placed for forty-eight hours in Erlicki's fluid which has the following composition :—

Potassium bichromate . . . .   2·5 parts
Sulphate of copper . . . . .   ·5 part
Water . . . .   100 parts.

This is done in order to free the preparations from the spirit, and they are then placed in the following mixture :—

Sulphate of copper . . . . .   ·5 part
Potassium bichromate . . . .   2·5 parts
Mucilage of syrup and gum . .   100 parts.

This solution with the specimens should be kept for two or three days in a warm chamber at 38° C.

Sections are then cut in the freezing microtome, received in Erlicki's fluid, washed in methylated spirit, and stained according to Weigert's method presently to be described.

If the specimens are cut with a Richert's microtome, sections will, as a rule, be easily obtained, but if the attempts prove unsuccessful, the precautions and measures suggested on p. 62 should be carefully followed.

For series-cutting the Cambridge Rocking microtome is undoubtedly the best, or series may be obtained by the process described on p. 64. It may so happen that after embedding in celloidin it is desired to cut series by the Rocking microtome. This may be done by removing the specimens from the corks, and immersing them in benzol, and then transferring to the liquid wax as described on p. 49. They may next be fixed to the Rocking microtome in the usual manner.

Staining.—Nearly all specimens of the central nervous system are now stained by one of three methods, namely, those of Weigert or Pal, or in aniline blue black.

For the demonstration of certain lesions the ordinary methods are preferable. For instance, when it is desired to examine miliary aneurisms in the brain, ammonium carmine should be used (see p. 78), or hæmatoxylin and eosin.

A simple method which stains the ganglion cells very well, is to dilute Stephens' blue-black ink one-half to two-thirds (Harris and Power).

As a rule sections are made for the purpose of investigating tracts of degeneration, and for this purpose one of the processes named above, and which we now proceed to describe, must be employed.

Weigert's method.—The solutions required are :—
Solution 1 :—

Saturated solution of acetate of copper . 1 part
Distilled water . . . . 1 ,,

Solution 2 :—

Hæmatoxylin . . . .   1 part
Alcohol (90 per cent.) . . . . . .   10 parts
Concentrated sol. of lithium carbonate   1 part
Distilled water . . . .   90 parts.

The hæmatoxylin must be dissolved in the absolute alcohol, the water added, and the mixture boiled. After it is cool the lithium carbonate is added. This solution will not be ready for use for a couple of days and will not keep for more than a month.

Solution 3 :—

Borax . . . .   2 parts
Ferricyanide of potassium . .   2·5 parts
Water . . . .   200 parts.

It is essential that the specimens should have been hardened in Müller's fluid.

The first step consists in placing the sections, whether single or in series, in the solution of acetate of copper for twenty-four hours, and then thoroughly washing them for about half that time in water, which should be frequently changed. This part of the process may be performed in mass before sections are cut, by placing the specimens after embedding in the copper solution for two or three days, or if a warm chamber at 38° C. can be employed, forty-eight hours will be sufficient, sections being then cut. When the washing is

complete the sections are placed in the hæmatoxylin solution for one or two days, and are then removed to water.
They should now exhibit a deep blue-black colour, and be
perfectly opaque. If this be not the case a drop or two of
a saturated solution of lithium carbonate may be added to
the water, and if the desired appearance is not then
produced the specimens must be returned to the hæmatoxylin solution for another twenty-four hours. They
must next be thoroughly washed in water which should
be changed every few hours, until the washings are no
longer coloured. Unless this process be thoroughly
carried out the sections will ultimately develop a number of dark spots which will render them practically
useless.

The next step consists in differentiating them. For
this purpose the sections are placed in the borax and
potassium ferricyanide solution (Sol. 3). They will at
once be seen to undergo a change, becoming lighter in
colour, and the grey matter gradually becoming separated from the white. This part of the process must be
carefully watched, and as soon as the two parts of the
cord are clearly defined they must be removed to water.
The time required varies from five minutes to half an
hour. If left too long the hæmatoxylin will be removed
and the specimens appear uniformly brown.

After washing in water the sections must be dehydrated, cleared in creosote, and mounted in balsam.
If the specimens were previously embedded in celloidin,
it will be better to use 95 per cent. alcohol for dehydrating, otherwise owing to the solution of the embedding material in the absolute alcohol, the sections will
probably fly in pieces, and thus all the labour and
trouble bestowed upon them will be lost.

Stained in this manner, the white substance of

Schwann assumes a purple-black colour, whilst the connective tissues, ganglion cells, and axis cylinders are dyed an orange-brown. If there are tracts of degeneration (sclerosis) these also assume the latter colour.

Pal's method.—This is a modification of the preceding.

The solutions required are :—

Solution 1 :—

Hæmatoxylin . . . . ·75 part

Distilled water . . . . 90 parts.

Alcohol . . . . 10 ,,

to which is added immediately before use a small quantity of a saturated solution of lithium carbonate in the proportion of three drops of the solution to ten c.c. of the hæmatoxylin stain.

Solution 2 :—

Permanganate of potash ·25 per cent. solution.

Solution 3 :—" Pal's solution."

Oxalic acid . . . . 1 part

Potassium sulphite . . . . . . 1 ,,

Distilled water . . . . 200 parts.

The specimens are hardened in Müller's fluid, as described above and embedded in celloidin, the sections being cut either singly or in series. They are at once transferred to the hæmatoxylin solution, where they remain for five or six hours, and are then transferred to water, containing a drop or two of a saturated solution of lithium carbonate.

They should then be bluish-black in colour and opaque.

The sections must remain in water, which should be frequently changed, until no more colour will come out.

The method of differentiating is the characteristic part of this process. The sections are first placed in the permanganate of potash solution for from fifteen to twenty seconds ; after which they are immersed in Pal's solution. Here they must be carefully watched until the white and grey matters are distinctly defined. This will probably be brought about in from one to two minutes. If any black spots appear the sections must be again placed in the permanganate of potash for a few seconds, and then again washed in Pal's solution.

After this they must be washed in water for a quarter of an hour. If they are now dehydrated (95 per cent. alcohol) cleared, and mounted in balsam, the medullary sheaths will be found stained a bluish-black colour, the remainder of the tissue being white.

Counter-staining is therefore advisable, and for this purpose borax carmine (see p. 79) is the best solution to use.

Pal-Exner method.—This process is more rapid than the preceding (although in some respects more troublesome), and produces very good results.

The nervous material cut into small segments is placed whilst fresh in about ten times its bulk of one half per cent. solution of osmic acid, where it remains two or three days, the solution being changed daily.

The specimens are then removed, washed in water, and dipped for a few seconds into absolute alcohol, and are then embedded in celloidin or paraffin.

Sections are cut in glycerine and removed to water. They are differentiated in the same manner as in Pal's solution, that is to say in permanganate of potash, and in the liquid known as Pal's solution (*vide supra*).

Aniline blue-black.—By staining sections of the central nervous system by this method, the nerve-cells are especially brought into prominence, being coloured a deep slaty blue. Two or three methods have been

introduced, but the best is undoubtedly the one suggested by Lewis (" Human Brain," p. 125).   Sections
of the fresh brain or spinal cord are cut as thin as
possible by means of an ether freezing microtome, are
then floated on water, and afterwards immersed for the
purpose of hardening in a two per cent. solution of
osmic acid.   After a few minutes (about eight to ten) they
are washed in water and placed in ·25 per cent. aqueous
solution of aniline blue-black for two hours.   The sections are once more washed in water, and each is spread
out on a slide, and allowed to become quite dry, after
which it is mounted by the addition of a drop of Canada
balsam and a cover-glass.

In all the above processes a number of changes are
required, and if the sections are moved from solution to
solution by means of needles, they are very apt to be
damaged, and much time is taken up in carefully manipulating them.   To avoid unnecessary risk, a small
perforated zinc sieve should be employed, in which the
sections may be placed, and so transferred altogether
from one solution to another.

If it be desired to prepare specimens of the cortex of
the brain, very good results may be obtained by adopting **Jolgi's "sublimate method."**   It consists of
two processes :—1. Hardening in bichromate of potash.
2. Treatment with bichloride of mercury.   Small pieces
of the tissue are placed in Müller's for about four weeks,
and are then passed directly into a watery solution of
corrosive sublimate.   The best strength to use is a ·25
per cent. solution.   The volume of fluid must be several
times that of the specimens, and it must be renewed as
often as it becomes yellow.   An immersion of at least
three weeks must be allowed, but longer than this improves the result.   Sections are then cut by the freezing

microtome, and afterwards thoroughly washed in water, otherwise they will be spoilt by the formation of a black precipitate. They are best preserved by mounting in glycerine. Prepared in this way the nerve cells and sometimes the blood-vessels are opaque and black. Occasionally the body of the cell and its finest processes are outlined with the utmost clearness.

This is not a true stain, but Jolgi thinks that there is formed in the tissue elements a precipitate of some substance that renders them opaque.

CHAPTER X.

Injection of Tissues.

The operation of injecting the blood-vessels is not often required for pathological purposes; but very instructive specimens of emphysematous lung may be prepared in this way. In the physiological laboratory it is more often performed. The process is rather a difficult one, and for successful results requires more complicated apparatus than a private individual generally cares to set up.

For the sake of completeness two simple methods will be described here, which with a little practice and patience will answer very well. Two kinds of "injection masses" are in use, namely, one which is fluid at ordinary temperatnres, and the other which is solid at the temperature of the air, but melts by heat; the latter is more generally used. Several varieties of each have been introduced, but only two are of importance :—

Cold injection.

Soluble Prussian blue . . .   2 parts.
Distilled water . . . . . . 100 ,,

The Prussian blue is dissolved in the water, and a few drops of hydrochloric or acetic acid are added before injecting.

Carmine gelatine injection mass.

Carmine . . . .   3 parts.
Strong ammonia . . . . 6 ,,
Glacial acetic acid . . . 6 ,,
Coignet's French gelatine . 7 ,,
Water . . . . 80 ,,

The gelatine is cut up into small pieces, placed in 50

parts of the water, and allowed to swell up for four or five hours. The carmine is ground up in a mortar with a little water, and the ammonia added. This mixture is allowed to stand for two hours, and then poured into a bottle, the mortar being rinsed with the remainder of the water. The swollen gelatine, with any remaining water unabsorbed by it, is placed on a water-bath until it melts. A dark purple fluid results from the carmine mixture, to which the acetic acid is added, a drop at a time, the whole being thoroughly mixed; as soon as the colour changes to a crimson the addition of the acid must be stopped. The crimson carmine is added little by little to the melted gelatine, the mass being continually stirred. Prof. Stirling states that this mass may be kept in a cool place for a long time if its surface be covered with methylated spirit. Before it is employed it must be melted over a water-bath, and filtered through fine flannel, rinsed out in hot water.

The organ to be injected should be removed immediately after death, the main artery being carefully dissected out and cut as long as possible. The injections may be made with a brass syringe, which should be short, and wide in the barrel to allow of its being readily cleaned, and it should have several nozzles of varying calibre, each provided with a shoulder and groove by which to secure it.

Instead of fixing a nozzle directly on to the syringe, it should be fastened into the vessel by means of thread, and a stout piece of elastic tubing should be fixed over it, connected by the other end to the syringe. When in use the nozzle is filled with the injection fluid, as is also the syringe with the tubing attached, the latter being closed at its extremity by a brass clip, so that all air may be excluded; the tubing is then applied to the nozzle, the clip removed and the piston slowly pushed in, the nozzle being steadied by an assistant. A pause must occasionally be made to allow the fluid to slowly force its way through the vessels. If the blue solution be used no further precautions are necessary, but if the carmine injection-mass be employed, the organ to be injected must be placed in warm water (40° C.), and before use the syringe must likewise be heated by drawing hot water up into it several times, the injection-mass itself being liquified over a water-bath.

More frequently injections are made by a "constant-pressure" apparatus. That described by Prof. Stirling ("Practical Histology," p. 75) is the simplest and best. This apparatus consists of a tin trough large enough to contain the organ to be injected, and sufficient water is poured in to cover it. The water is kept at 40° C. by means of a gas burner or spirit lamp placed beneath it. In the same trough is placed the injection-mass in a Wolff's bottle. The bottle is connected with a large air chamber (a large wide-mouthed jar answers very well) into which water can flow from the tap, and thus compress the air. The pressure within the chamber can, if desired, be registered by means of a manometer. The compressed air acts on the surface of the injection-mass in the Wolff's bottle, and forces it through a tube which is attached to a cannula fixed in the main vessel of the organ.

After the operation is over, the further treatment depends upon which material was used. If the Prussian blue solution has been employed, the organ must be at once placed in equal parts of methylated spirit and water to which a few drops of hydrochloric acid have been added; it is left in this for twenty-four hours, after which it may be cut up and hardened in alcohol.

When the warm carmine mass has been employed, the organ is rapidly cooled by being placed in running water until the mass is thoroughly set, when it may be cut into small pieces and hardened in alcohol.

CHAPTER XI.

EXAMINATION OF FRESH TISSUES.

IT may be often necessary, either in the post-mortem room, or especially in the operating theatre, to examine a tissue as soon as it is removed from the body for purposes of diagnosis. Although the specimens so prepared are not as satisfactory as when they have been submitted to the processes of hardening, and the other stages necessary for mounting permanently, yet by means of certain precautions very fair results may be obtained.

Tissues cannot be examined in a dry state on account of their opacity and the irregularity with which they transmit light. They must, therefore, be bathed in some medium which will change their appearances and vital properties as little as possible. The examination of blood, pus, effusions, &c., will be referred to in future chapters, and we shall consider here the treatment of tumours, and the method of examining such tissues as nerve fibres, splenic pulp, membranes, &c. The fluids usually employed for the purpose of moistening such specimens are, normal saline solution, serous fluid, aqueous humour, and oxidised serum.

Normal saline solution.—The preparation of this liquid was given on p. 33. It can be prepared in bulk and kept continually at hand. It alters the tissue but slightly, and is the most convenient of all these liquids.

Serous fluid.—This is most commonly taken advantage of in the post-mortem room where it may always be procured from the pericardial sac.

Aqueous humour.—This is obtained by puncturing the anterior chamber of the eye of a recently killed ox.

Iodised serum.—The usual formula for this is—

Tincture of iodine . . .   1 part
Serous fluid . . . . .   100 parts
Carbolic acid . . . . .   .5 part.

Stirling ("Practical Histology," p. 24) recommends that a strong solution should be kept which may be diluted as required. The serous fluid obtained by allowing blood to coagulate, and pouring off the serum from the clot is placed in a bottle with a few crystals of iodine. This mixture must be frequently shaken. At first very little iodine is dissolved, but in a fortnight or three weeks the solutions become of a deep brown tint. When required, a little of the strong fluid is added to fresh serum until the latter has a light brown colour. In whatever manner prepared, this fluid alters the tissues slightly, and colours them a light yellow.

Thin **membranes** may be placed at once on the slide, spread out as far as possible by means of scissors and forceps in a few drops of normal saline solution, and a cover-glass applied. A better result is obtained if a drop of a dilute watery solution of methylene blue is added. The stain must be very weak, or it adds to rather than detracts from the obscurity of the specimen.

In the examination of fresh tissues, as a rule, the operation of **teasing** has to be adopted. A small portion of the muscle or other fibrous material is cut off with a pair of curved scissors and placed on a clean glass slide with a small quantity of one of the solutions previously named. The elementary parts of the tissue are then separated by means of needles. To effect this the particle under examination is fixed with one needle, whilst small fragments are torn off with the other. These are treated in the same manner until they are as minute as possible. The most suitable fragment is then placed beneath a cover-glass, the bulk of the tissue being removed.

Another method which may sometimes be adopted, is to scrape the surface of the object to be examined, having previously moistened it with saline solution, and the scrapings are collected on the slide and examined under a cover-glass.

Great assistance in either of the above processes is obtained by **"softening"** or **"dissociation."** Several reagents are used for this purpose, of which the following are the best :—

Iodised serum prepared as above may be used for isolating nerve fibres. Specimens must be allowed to remain in it for thirty-six hours, and the strong iodised serum should be diluted until the solution is of a deep cherry colour.

A ten per cent. solution of **common salt** may be employed to soften connective tissue, being especially useful in the examination of tumours.

Dilute alcohol, according to Ranvier, is very serviceable for epithelial tissues ; twenty-four to thirty-six hours is required for its full action. The best strength to use is one part of 90 per cent. alcohol with two parts of water.

Caustic potash in 40 per cent. solution is recommended by Woodhead for isolating muscle cells. This is very rapidly accomplished, seldom requiring longer than half an hour.

Potassium bichromate in two per cent. solu-

tion is useful for dissolving the cementing material between the fibres of tendon, or for dissociating epitheliun,
and the nerve cells of the spinal cord. It requires
seven to fourteen days to complete the process.

For specimens of the central nervous system,
Landois' fluid is perhaps as good as any. It has
the following composition :—

Sodium sulphate . . . . .   5 parts.
Neutral ammonium chromate .   5 ,,
Potassium phosphate . . .   5 ,,
Distilled water . . . . . .   100 ,,

Small pieces of tissue must remain in it from one to
five days:

In the investigation of tumours or any solid tissue for
diagnostic purposes, it is advisable to attempt to procure sections. After a considerable amount of practice small sections may be cut with a sharp razor.
The specimen is held in the left hand and short quick
cuts are made across it, the blade of the razor then
being dipped in a glass capsule containing water. The
vessel should be placed over a black surface, and the
cut particles gently moved about with a needle, two or
three sections will then generally be found thin enough
for mounting.

A very useful little piece of apparatus for similar purposes is Valentin's knife (see fig. 18) which consists of
two parallel blades fixed in a handle which can be approached to or separated from one another by means of
a screw. The tissue to be cut is held in the left hand.
The blades of the knife are then set almost close to
one another, moistened with water and drawn with
a clean cut through the specimen, the resulting section
being separated by a sharp turn of the instrument, and
the blades afterwards separated under water. Although

this operation appears very simple it requires consider-able practice before satisfactory sections will be ob-tained.   The chief difficulty lies in judging the distance at which the blades should be set from one another for various substances.   Experience alone can teach this accurately.

The most satisfactory method of obtaining sections of fresh tissues is to dip them in gum (see p. 48) and to cut the sections by means of an ether freezing microtome. The manner of carrying this out has already been de-scribed on pp. 56 and 58.   The sections may afterwards be stained, watery solutions of the dyes being preferable. Bismarck brown, magenta and methylene blue may be used, and the specimens afterwards mounted in gly-cerine.

CHAPTER XII.

PREPARATION OF INDIVIDUAL TISSUES AND ORGANS.

IN commencing the study of the microscope in medicine some difficulty will be probably experienced as to which of the foregoing methods is applicable to particular tissues and organs. Some hints in regard to this matter will be found distributed through the previous chapters, but for the sake of easy reference directions will be given in this chapter to assist beginners in their choice of hardening and staining reagents gathered from the personal experience of the author. But after some practice the student will be competent to select his own favourite and successful processes.

It should be borne in mind that in order to fully understand diseased organs and tissues the normal structnre of these parts must be thoroughly known. The student will do well therefore, to examine a number of healthy organs obtained from the post-mortem room, or from small animals such as cats, guinea pigs, &c., and in the preparation of these considerable proficiency in hardening and staining will be acquired.

We shall begin the list with normal histological structures, such as epithelium and then pass on to the various organs.

Squamous epithelium.—This occurs either in a single layer, or in several layers; in the latter case it is said to be stratified. As a single layer it lines serous membranes: blood and lymphatic vessels, &c., and is known as endothelium. A characteristic specimen may be obtained by killing some small animal such as a guinea pig or cat, and immediately opening the abdomen, removing portions of the omentum, or the mesentery, and spreading them out on cork. They should then be placed at once in a solution of silver nitrate (p. 87), so as to demonstrate the outlines of the cells. Other specimens should be stained in hæmatoxylin or alum carmine. Stratified squamous epithelium may be obtained from sections of skin, or better from the tongue of any small animal. They may be hardened in absolute alcohol, embedded in celloidin, and stained in picro-carmine or hardened in Müller's fluid, cut by freezing and stained in carmine.

For the purposes of demonstrating karyokinesis, the specimens should be stained in safranin or better, submitted to the method described on p. 103.

Columnar epithelium.—This form of epithelium lines the mucous membrane of the alimentary canal from the cardiac orifice of the stomach, downwards, and the greater part of the ducts of the glands opening into it. Isolated cells may be obtained by scraping the mucous membrane of the intestine of some animal, or hardening a portion of the intestine in Müller's fluid, embedding in celloidin and staining in hæmatoxylin and eosin.

Secretory epithelium.—Typical examples of this variety of epithelium may be obtained by scraping the cut surface of a liver and examining in glycerine, or by cutting sections of a liver, and staining in hæmatoxylin and eosin.

Ciliated epithelium.—This form is best obtained from the trachea of any small animal, hardening the specimen in Müller's fluid, embedding in paraffin and staining in hæmatoxylin. Very satisfactory specimens may be procured from an ordinary nasal polypus by hardening immediately after removal in a saturated watery solution of corrosive sublimate for two or three hours, then thoroughly washing it in alcohol, and staining in hæmatoxylin.

Transitional epithelium.—This type of epithelium is best seen in the bladder ; to obtain specimens the bladder should be spread out on cork, and placed in dilute alcohol for twenty-four hours, and stained *en masse* in picro-carmine. A little of the mucous surface is scraped off and examined in glycerine.

Areolar tissue.—This is rather difficult to obtain. The best method is that described by Harris and Power (" Physiological Laboratory", p. 92), " By the injection into the subcutaneous tissue of a rat which has just been killed, of a 0·2 per cent. solution of nitrate of silver or osmic acid, a small artificial bulla is formed. This is allowed to remain for from ten to thirty minutes and is then opened with a pair of fine curved scissors, and the delicate subcutaneous tissue is rapidly removed and spread out on a glass slide. It is immediately covered with a cover-glass, and the preparation is stained for 24 hours with picro-carmine. Glycerine is passed through until all the superfluous staining material is removed after which the preparation is sealed up."

White fibrous tissue.—Typical examples of this tissue are found in tendons. Samples may be obtained either from the tail of a mouse or rat, or from an amputated limb. Preparations may be made by teasing in normal salt solution, and staining with magenta, or portions may be hardened in Müller's fluid, cut by freezing and stained in hæmatoxylin and eosin.

Elastic tissue.—This is most readily demonstrated by teasing out a small piece of the ligamentum nuchæ of an ox in glycerine, or by mounting a piece of the omentum of some small animal, washing it freely with dilute acetic acid, staining with hæmatoxylin and mounting in glycerine.

Adenoid tissue (retiform).— This is best exemplified by hardening a lymphatic gland in Müller's fluid, cutting by freezing, and staining in hæmatoxylin. Some of the sections should then be mounted directly, and others placed in a test-tube half filled with water, and thoroughly shaken. This dislodges the lymph corpuscles, and leaves the fine reticulum visible. The sections may then be dehydrated, cleared, and mounted in balsam.

Adipose tissue.—For this purpose specimens of skin hardened in Müller's fluid, cut by freezing and stained in hæmatoxylin should be prepared, whilst other portions should be treated at once with osmic acid, so that the action of this reagent on fat cells may be observed.

Gelatinous or embryonal tissue.—This form of tissue occurs typically in the umbilical cord, and in the skin of the embryo. A portion of the cord may be hardened in Müller's fluid and then in alcohol. Sections are cut by freezing, and stained in picro-carmine. Very beautiful specimens may also be obtained by forming a bulla by the injection of a dilute solution of gold chloride into the subcutaneous tissue of an embryo, in a stronger solution of which it is subsequently stained (Harris and Power).

Cartilage.—There are three chief varieties, hyaline, elastic, and fibro-cartilage.

Hyaline cartilage.—Specimens may be obtained from the ribbed cartilages of young animals. They should be hardened in chromic acid or spirit, and sections cut by freezing, and stained in hæmatoxylin and eosin.

Elastic cartilage obtained from the epiglottis or cartilages of the ear, should be hardened in spirit and stained in picro-carmine.

Fibro-cartilage.—This variety occurs in the intervertebral substance. Preparations may be hardened in spirit, and stained in hæmatoxylin and eosin. Other sections should be stained in chloride of gold.

Ossifying cartilage.—A long bone of a fœtus, such as the femur, is decalcified in Perenzi's solution (see p. 45). Transverse and longitudinal sections are cut in celloidin, and stained in hæmatoxylin and eosin.

Bone.—Specimens are decalcified in Perenzi's fluid, cut in celloidin and stained in picro-carmine.

Teeth.—Specimens are placed in a saturated solution of picric acid until quite soft, then hardened in alcohol, cut in celloidin, and stained in picro-carmine or hæmatoxylin and eosin.

In order to show the development of teeth, the jaw of a newly born kitten or puppy should be obtained, and first placed in Von Ebner's solution, then hardened in alcohol, and stained in lithium carmine and picric acid.

Striped muscle.—Any voluntary muscle may be taken, hardened in chromic acid and spirit, cut in celloidin and stained in picro-carmine. Fresh sections should also be examined by teasing in glycerine so as to make oneself familiar with unstained muscular fibres such as are often met with in sputum and vomited matters.

Unstriped muscle.—The most satisfactory method of demonstrating this tissue is by hardening portions of intestine or stomach in Müller's fluid, cutting in celloidin, and staining in hæmatoxylin and eosin.

Central nervous system (see special chapter).

Brain.—Specimens must be hardened in Müller's fluid or in ammonium bichromate, and cut in celloidin. Sections should be stained in various ways, in carmine, aniline blue-black, and by the methods of Weigert and Pal.

Spinal cord.—The same methods as for the brain may be employed. If there is reason to suspect degeneration, Weigert or Pal's method should certainly be chosen.

Nerve fibres.—Portions of nerves should be taken from recently killed animals; some of them may be hardened in Müller's fluid, cut in celloidin, and stained in osmic acid, picrocarmine, or by Weigert's method.

Others should be placed at once in osmic acid, then teased out, and mounted in Farrants' solution.

Nerve cells.—In order to demonstrate the different variety of nerve cells, specimens of the following should be hardened in Müller's fluid, cut in celloidin and stained in aniline blue-black.

1. Spinal ganglia.
2. Sympathetic ganglia.
3. Spinal cord to exhibit multipolar nerve cells.
4. Cerebrum.
5. Cerebellum.

Nerve endings. Simple tactile cells.—These can be well demonstrated in a section of skin, placed immediately after removal (*e.g.*, after amputation) in a solution of chloride of gold. Sections are then cut by freezing, and mounted in Farrants' solution.

Pacinian corpuscles.—To display these bodies, sections of skin should be cut, preferable from the fœtus, hardened in alcohol, and stained in hæmatoxylin. (For further information as regards nerve-endings, see " Outlines of Practical Histology," Lesson xxxiv., by Prof. Stirling).

Blood-vessels. Arteries.—The structure of an artery is best illustrated by hardening a piece of the aorta or carotid artery of a cat in Müller's fluid, cutting in celloidin and staining in picro-carmine.

Veins.—These may be prepared in a similar manner.

Capillaries.—These are best obtained from the pia-mater. The brain of an animal is placed for two days in a 2 per cent. solution of bichromate of potassium. The pia-mater is then stripped off, stained in hæma·toxylin and eosin, and mounted in Farrants' solution.

Lymphatic glands.—The lymphatic gland of any animal may be hardened in Müller's fluid, cut in celloi-din, and stained in lithium carmine and picric acid.

Specimens of **sweat** and **sebaceous glands** may be prepared in the same way ; the former being taken from the sole of the foot, and the latter from the skin of some animal.

The eye.—The structure of this organ may be demonstrated on the eye of a bullock which has been recently killed. It should be hardened in Müller's fluid for three weeks, embedded in celloidin, and the sections stained in hæmatoxylin and eosin. (For a full account of the pathological examination of the eyeball see a " Manual of Clinical and Practical Pathology," by Drs. Wynter and Wethered).

Internal ear.—The temporal bone of a cat or dog is removed and decalcified, and afterwards hardened in methylated spirit. It is then embedded in celloidin, and sections cut in the direction of the longitudinal axis of the cochlea.

Tongue.—Specimens of this organ should be hard-ened in Müller's fluid, cut in celloidin, or by freezing, and stained in safranin and in picro-carmine.

Specimens of **œsophagus, stomach,** and **intes-**

L

tine should be hardened first in Müller's fluid, and then in absolute alcohol. After embedding in celloidin, the preparations should be so arranged that sections are cut transversely and longitudinally to the axis of the intestines. Hæmatoxylin and eosin is the best staining method. For the purposes of instruction, sections of the stomach should be made through the cardiac and pyloric ends, and from a portion of the greater curvature. From the intestine, portions should be selected from the upper part of the duodenum, the ileum, and colon.

Lungs.—Portions should be hardened in absolute alcohol, cut in celloidin, and stained in lithium carmine and picric acid. The student should make himself familiar with the structure of the bronchi and trachea. The process of embedding in celloidin is especially useful in examining specimens of lungs, more particularly in cases of pneumonia and phthisis. It is almost essential when micro-organisms (*e.g.*, tubercle bacilli) are to be sought for.

Thyroid gland.—Specimens are to be prepared by hardening the gland for twenty-four hours in a mixture of spirit and water, and then in absolute alcohol. Sections should be cut in celloidin and stained in picrocarmine.

Thymus gland removed from a fœtus should be treated in the same way.

Liver.—Hardened in absolute alcohol, cut by freezing, and stained in hæmatoxylin and eosin.

Sections of kidney are prepared in the same manner.

Pancreas and **supra-renal bodies.**—Hardened in Müller's fluid, cut in paraffin or celloidin, and stained in hæmatoxylin and eosin.

Spleen.—Harden in Müller's fluid. Place a few sections in a test-tube with water and thoroughly shake; then mount the sections unstained so as to show the structure of the pulp. Other sections should be cut in celloidin, and stained in carmine and picric acid.

Bladder.—The organ should be distended with chromic acid and spirit, and immersed in the same mixture for twenty-four hours. Portions are then cut off and hardened in absolute alcohol, and treated in the same way as the intestine.

Portions of the **uterus, ovaries,** and **Fallopian tubes** should be hardened in absolute alcohol, cut in celloidin, and stained in hæmatoxylin and eosin. Specimens of **testicle** may also be treated in like manner.

Embryological specimens should be hardened in Müller's fluid, cut in celloidin, and stained in hæmatoxylin and eosin.

CHAPTER XIII.

THE EXAMINATION OF TUMOURS.

ONE of the most frequent tasks which a pathologist is requested to undertake is the microscopic examination of tumours, and as this subject, therefore, comes strictly under the head of Medical Microscopy, it has been considered advisable to devote a chapter to a brief description of the most important growths of this kind. In the operating theatre, sections may be made, directly the tumour is removed, by means of a Valentin's knife, and examined either unstained in salt solution (see p. 33), or preferably stained by dipping them into a solution of hæmatoxylin. Their nature can then be generally easily recognized. Better specimens can be obtained in a very short time by cutting off a small piece of the tumour, placing it for a few minutes in gum solution, freezing it on a Cathcart's or Swift's microtome, and so obtaining thin and complete sections. If permanent preparations are required, the tissue must be hardened, stained, and mounted by the processes already described. The following classifications and descriptions are largely taken from Mr. Frederic S. Eve's chapter on tumours, in Mr. Mansell Moullin's "Surgery."

Although cysts are usually described with tumours, an examination of their contents will be considered in a future chapter.

Fibromata.—These are tumours composed of adult or well-developed fibrous tissue. In consistence they

vary from a density approaching that of cartilage, to
a succulent, yielding, but still not friable tissue.

A section of the firmer fibromata displays a number
of fibrous bands standing out on a grey, yellowish,
or white ground; or the section is firmly uniform and
dull white.   The fibres either interlace, or are arranged
concentrically or in parallel lamellæ.   A scraping yields
no juice.   Microscopically, the firmer tumours are composed of looser or more compact bundles of fibres, often
wavy and interlacing or disposed chiefly parallel to each
other.   Situated on the bundles in small numbers, are
flattened nuclei, belonging to connective tissue cells,
the protoplasm of which may be shown to anasto-
mose by processes around the bundles.   The softer
forms are made up of interlacing fibrillæ, more or less
thickly studded with large round or oval connective
tissue cells, whose protoplasm is continuous with the
fibrillæ.   The chief diagnostic characteristics are then,
fibrous tissue with scanty cells, and at growing points
young cells surrounding adult and well developed
fibrous tissue.

Myxomata (fig. 20).—Myxomata are new formations
of loose, fibrillar, connective tissue, permeated with
fluid which is rich in mucin.   Common examples are
polypi of the nose, rectum and uterus.

A section exhibits a glistening, pale, jelly-like tissue,
the surface of which is raised in the centre.   Microscopically the tumour is composed of stellate and
round cells, of which the protoplasm is prolonged in
delicate filaments which meet one another and form a
felting of fibrils.   The stellate cells are chiefly characteristic of mucous connective tissue; in some growths
they exist exclusively, while in others the round cells
preponderate.   The vessels are usually abundant and

clearly seen, owing to the transparency of the tissue; they often form a wide meshwork.

Enchondromata (fig. 21).—These are tumours composed of cartilage chiefly of the hyaline variety. The section is glistening, smooth and translucent, and almost invariably shows a number of separate masses

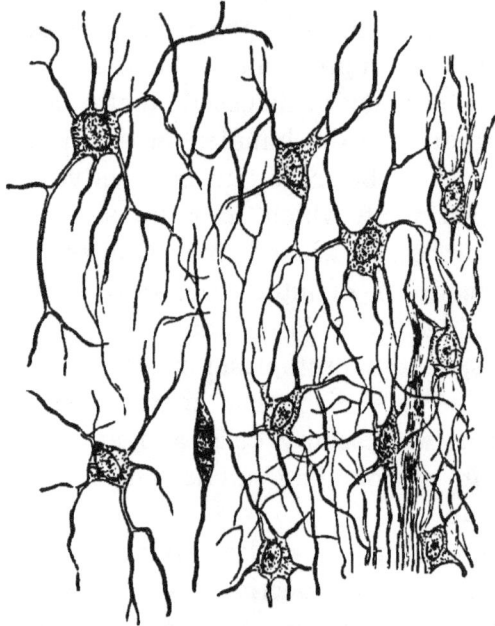

FIG. 20.—Cells from a periosteal myxoma of the thigh (Ziegler).*

or lobes divided by bands of connective tissue containing blood-vessels. In some cases the section is often greatly modified by secondary changes.

Microscopically, the commoner forms have practically the same structure as ordinary hyaline cartilage. In many the cartilage cells are more irregularly distributed

* Figures 20, 21, 22, 23, 25, 26, from Ziegler's "Pathological Anatomy" are inserted with the kind permission of Messrs. Macmillan and Co.

and more numerous in proportion to the matrix; the
capsules are larger, and in growing parts of tumours a
single capsule often contains two or more cells. The

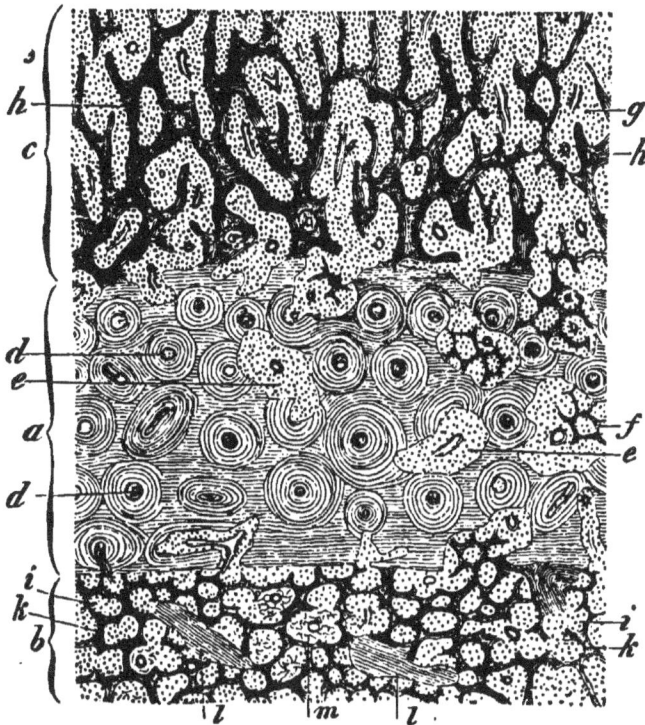

. FIG. 21.—Section from an osteoid chondroma of the humerus
(Ziegler). *a*, cortical layer; *b*, medullary spaces or cancelli; *c*,
periosteal growth; *d*, normal Haversian canals; *e*, Haversian canals
distended with cartilage, which at *f* contains a core of new bone; *g*,
cartilage developed from periosteum which at *h* contains bony tra-
beculæ; *i*, cartilage developed from medullary tissue, which at *k*
contains bony trabecula; *l*, original trabecula; *m*, remnants of
medullary tissue.

cells at the periphery beneath the perichondrium are
flattened. Not rarely stellate ones are observed, especially in the parotid tumours.

Osteomata.—Of these tumours there are three varieties; the eburnated, compact and cancellous. The eburnated occur as small flat elevations on the surface of the skull. Similar growths also spring from the bones of the face.

Compact osteomata are observed on the shafts of the long bones, with which their structure is identical.

The name cancellous sufficiently describes the struc-

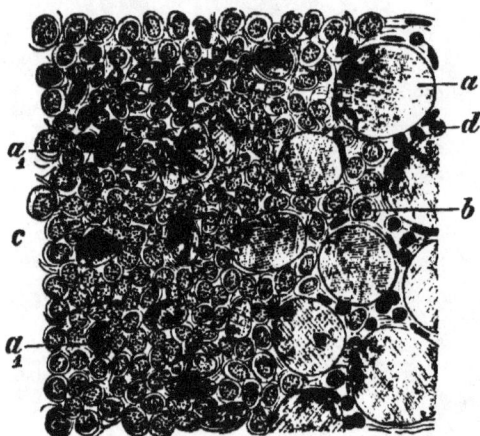

Fig. 22.—Section through the margin of a sarcoma affecting the intermuscular connective tissue (Ziegler). *a*, normal muscle fibre; *a₁*, atrophied muscle fibre; *b*, round cells intruded between the muscle fibres; *c*, fully developed tumour tissue; *d*, round cells resembling white blood corpuscles.

ture of the third variety, the most common and widely distributed of all. With rare exceptions they are situated on the diaphyses of the long bones, near to but not necessarily over the epiphysial disc; they are covered with a layer of cartilage, the deeper surface of which grows continually and is gradually converted into bone.

It is of course impossible to examine these tumours

by the methods described at the beginning of this chapter. They are, however, more easily recognised by their grosser characteristics, their position, &c., than by microscopical examination.

In order to procure sections, it is necessary to place the portions which have been removed in one of the de-calcifying solutions previously described until all lime salts have been removed, when sections may be cut in the ordinary way.

Myo-fibromata.—As the name implies these tu-mours are composed of muscle and fibrous tissue in varying proportions. The muscle presents the ordinary characters of the unstriped, or involuntary variety, and is distributed in inter-lacing fasciculi. The outline of individual fibres is often not easily distinguishable in sections, but the fibres may be readily isolated by the process of teasing as described on p. 135, chromic acid being the best medium in which to perform the opera-tion. The naked eye characters resemble those of a coarse fibroma, the sections being marked by distinct interlacing fibrous bundles.

Sarcomata.—This large class of tumours consists of growths derived from the connective tissue constitu-ents of the body, but their elements never develope into fully formed or adult structures. The sarcomata are divided into three chief varieties, the round-celled, spindle-celled and myeloid. Between these many inter-mediate forms are observed ; thus, round and spindle-cells may be mingled in nearly equal proportions—mixed sarcoma ; and myeloid cells are associated either with round or with spindle-cells. Again the spindle-celled sarcomata shade off indefinitely into the fibromata.

Round-celled sarcomata (fig. 22).—These again are divided into two sub-varieties, the small and the

large.   The former is much the commoner, and its
component cells are about the size of leucocytes or
granulation cells.   In the latter they are as large as
those of squamous epithelium.   A section of a round-
celled sarcoma under the microscope shows an uniform
surface formed of closely crowded cells, intersected here
and there by a band of connective tissue.   In this respect and in the character of the cells it differs essentially
from cancer.   The nuclei are large and the protoplasm

FIG. 23.—Section from a giant-celled sarcoma showing also
spindle-shaped cells (Ziegler).

around them scanty, in fact many of the small round-
celled tumours appear to be composed entirely of nuclei
and if the processes described on p. 103 are applied to
the sections karyokinesis will be demonstrated.   The
inter-cellular substance varies ; it is usually scanty and
homogeneous, sometimes formed of delicate interlacing
fibrillæ prolonged from the protoplasm of the cells, and
sometimes of homogeneous bands.   The blood-vessels

are usually abundant, and in many instances have no proper wall.

Spindle-celled sarcomata (fig. 23), are composed of either small or large spindle cells analogous to the fibroblasts of young granulation tissue ; they possess an oval nucleus, and their extremities are prolonged into a more or less delicate fibre. The cells are closely approximated, and are arranged in parallel or intersecting fasciculi or bundles. The protoplasm of the cells unites to form a coarse stroma.

Myeloid sarcomata (fig. 23) occur almost exclusively as tumours in the centre of bones, and in the gum as epulis. They are characterised by the presence of "giant cells" which consist of irregular masses of protoplasm containing many nuclei resembling those met with in tubercle. These cells are associated with others, either round or spindle shaped, or both. A section has, in parts, or over the whole surface, a peculiar maroon-red tint, by which the nature of the tumour may be recognised.

Sarcomata connected with bone often undergo partial transformation into cartilaginous or osseous material. In the former (chondro-sarcoma) the intercellular substance is abundant and hyaline. The osteo-sarcomata are composed either of round or spindle cells. The lime salts are deposited in a delicate fibrous reticulum in which the round cells are enclosed.

Some other varieties of sarcomata have to be mentioned :—

Myxo-sarcomata.—These occur especially in the breast, parotid gland and testicle. In microscopical sections they exhibit, in varying proportions, branched and round cells, which stand out distinctly in an abundant homogeneous stroma.

Alveolar sarcomata.—This variety is much more uncommon. It occurs in bone muscle, and subcutaneous tissue. The cells are large, round, and not unlike epithelial tissue. They are collected into alveolar masses bounded by delicate bands of fibrous tissue.

Melanotic sarcomata.—These originate in moles of the skin and in the choroid of the eye. They have a blackish colour, but the pigment is often distributed in irregular masses. In sections they exhibit an alveolar structure with a scanty stroma. The cells are usually round and large in size; occasionally spindle cells are present.

The pigment is chiefly deposited within the cells in the form of small granules, but some of it will be seen to lie free in the stroma.

Lympho-sarcomata.—This is a variety of round-celled sarcoma originating in a lymphatic gland, but never involving simultaneously the whole lymphatic system. Microscopically round cells are seen arranged in irregular cylinders lying in the meshes of a coarse fibrous stroma, but no definite reticulum can be discovered.

Adeno-sarcomata.—This can scarcely be called a distinct form of tumour. It consists of sarcomatous tissue generally of the round-celled variety, enclosing portions of the gland tissue in which it is situated; thus it is frequently seen in the breast, parotid glands and testicle, but the gland tissue must not be mistaken for new growth.

EPITHELIAL TISSUE TUMOURS.

This is a large and important group, and consists of tumours derived from the epithelium of the surfaces and glands of the body. There are two main divisions, the innocent, comprising papilloma and adenoma, and the malignant, *i.e.*, carcinoma.

Papillomata.—In the simplest form these are merely overgrowths of the normal papillæ of the skin or mucous membrane. Others are composed of a series of mesentery outgrowths, and some occur as branched

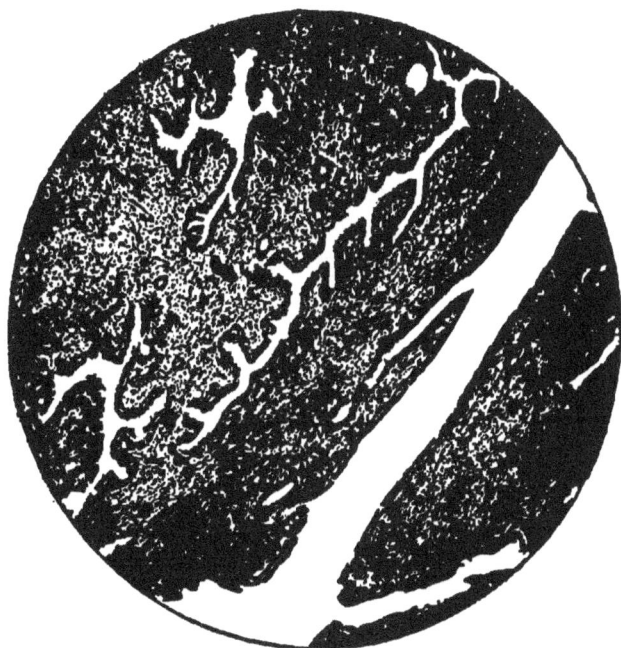

FIG. 24.—Fibro-adenoma of mamma (Boyce).

masses which are seen under the microscope to be covered with columnar epithelium (villous growths). The external form varies greatly, but in microscopic sections they are seen to consist of layers of epithelial cells mainly of the pavement variety, more or less altered in shape from pressure.

Adenomata (fig. 24).—These are glandular tumours

and differ in structure according to the gland which they attack. The most common variety is perhaps the mucous polypi which are composed of hypertrophied mucous glands. In microscopic sections branching tubules are seen which are lined with a single layer of columnar epithelium. The stroma is soft and made up of fibrillar tissue with stellate and round cells. Mr. F. S. Eve (*l. c.*, p. 135) says that the statement that adenomata do not secrete is erroneous, for the tubules of rectal polypi pour out quantities of mucus ; in one specimen of fibro-adenoma of the breast he found a quantity of milk which exuded from the section.

Carcinomata.—Cancers are tumours whose structure follows the type of the epithelial tissues.

Anatomically they are divided into groups in accordance with the nature of their epithelium.

Squamous-celled epitheliomata (fig. 25).— These occur in situations where there is normally a flat epithelium, namely, the skin, mouth, pharynx, œsophagus, larynx, vagina, &c. In some of the cases the epithelial processes tend to penetrate into the underlying structures so as to produce a more or less diffused infiltration and thickening of the part. As they grow, these processes soon become compressed, producing concentric whorls or small epidermic globes which are known as cell-nests or bird's-nests. The cells as a rule are large, and generally irregular in shape owing to pressure.

Rodent ulcer or cancroid of the skin is also a slowly growing form of epithelioma affecting chiefly the face of people advanced in years. Under the microscope it is seen to be made up of round or lobulated masses of very small, round, or slightly flattened cells. Cell-nests are absent as a rule.

Cylindrical-celled epitheliomata.—This affects parts where the surface is normally covered with cylindrical epithelium, hence it is chiefly found in the stomach and intestine. Microscopically, the tumour

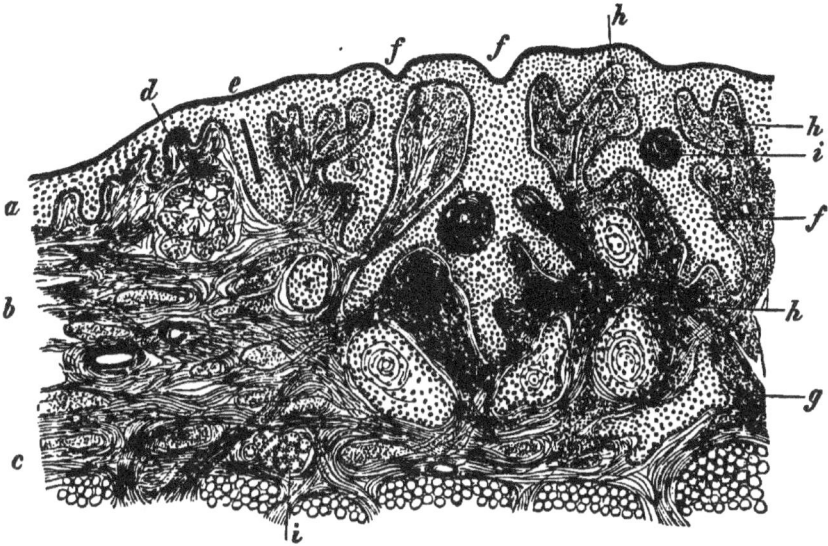

FIG. 25.—Section from an epithelioma of skin (Ziegler). *a*. epidermis; *b*, corium; *c*, subcutaneous areolar tissue; *d*, sebaceous gland; *e*, hair follicle; *f*, cancerous ingrowths from the epidermis; *g*, deep set cancerous cell groups; *h*, proliferating fibrous tissue; *i*, (above) cell nest; *i*, (below) sweat gland.

will be seen to be composed of a number of cylinders and alveoli lined with cylindrical epithelium. The stroma is well marked.

Spheroidal-celled cancer.—This is the most common form of carcinoma. It used to be divided into two varieties, hard or scirrhus, and soft or medullary, but these terms have lost their pathological significance, the difference depending only on the relative amount of stroma and epithelial cells respectively.

In the hard variety (scirrhus, fig. 26), the cells are small and arranged in elongated groups; they vary greatly in shape and often contain two nuclei.   Fibrous tissue is abundant and dense so that the walls of the alveoli are broad and well marked.   The cut surface of

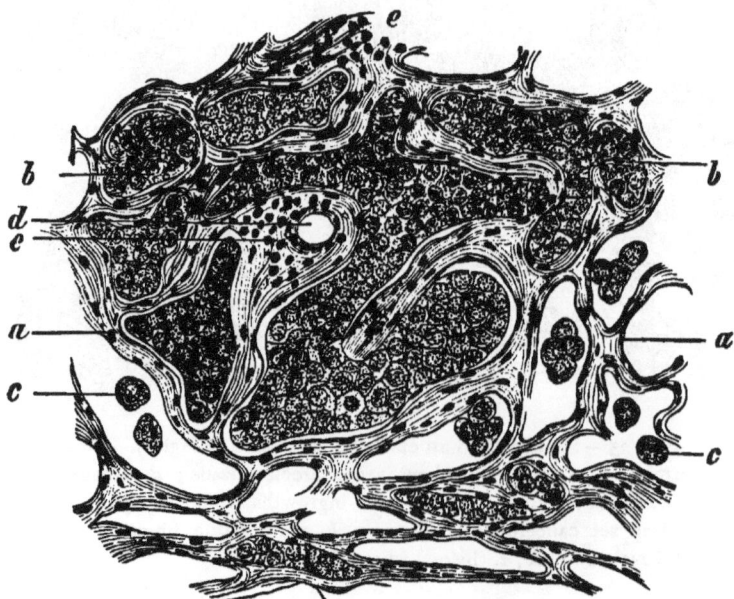

FIG. 26.—Carcinoma of the mamma (Ziegler). *a*, stroma; *b*, nests; *c*, cancer cells; *d*, blood-vessel ; *e*, fibrous stroma infiltrated with small cells.

the tumour is greyish and transparent, with opaque yellow markings indicating the existence of fatty degeneration in the cells.   But little juice can be obtained from the cut surface.   When placed under the microscope it is seen to contain irregularly-shaped cells.

In the softer tumours (medullary cancers) the stroma

is scanty and delicate. The cells are large, often coalescent and arranged in long columns or alveoli. The cut surface of the tumour is grey in colour, and juice can be freely scraped from it. In this juice, cells and free nuclei may be found, the latter being large, and mainly oval in shape. Carcinomata undergo fatty mucoid and colloid metamorphosis. The fatty change is found in most cancers, especially those of the breast. When the mucoid degeneration has advanced (**mucous cancer**) the tumour as a whole is gelatinous in appearance. The change is chiefly found in the stroma, the alveoli being bounded by broad transparent hyaline bands of connective tissue. A section presents a uniform, moderately firm gelatinous aspect.

Colloid cancer is found in growths of the stomach and intestines. The epithelium is chiefly affected by the metamorphosis, and the stroma only to a lesser degree. Small drops of colloid material make their appearance in the protoplasm of the epithelium, pressing the nucleus to one side, and distending the cell in the form of a vesicle. The cell wall gives way, and the cells merge together in a gelatinous mass ; to the naked eye such tumours have a markedly gelatinous appearance, and on section exhibit large alveoli bounded by thin bands of stroma and containing soft white jelly-like material. These tumours frequently occur as infiltrations penetrating among the constituents of the tissues.

If cancer attacks the skin or eye-ball it may become pigmented and is then known as **melanotic carcinoma,** but it is a rare form of tumour compared with melanotic sarcoma.

M

CHAPTER XIV.

EXAMINATION OF URINARY DEPOSITS.

IN this chapter will be described the microscopic examination of urinary deposits.   The chemical examination, which is even of greater importance, cannot be
considered here.   The reader is referred to larger works,
notably " Clinical Diagnosis " by Prof. v. Jaksch (translated by Dr. Cagney) and " Urinary and Renal Diseases " by Sir William Roberts.

Many morbid conditions of the urinary tract, however, are characterised by a particular deposit, and it is
desirable that whenever disease of the kidney, bladder,
urethra or genital tract is suspected, that a microscopic
examination of the urine should be made as well as a
chemical ; this is especially to be insisted on whenever
there is a trace, no matter how small, of albumen in the
urine, and also when there is any suspicion of the formation of a calculus.

The urine should be collected in a conical glass and
allowed to stand carefully protected from dust for about
twenty-four hours, by which time the deposit will have
fairly settled.   It is advisable to add some disinfectant
(*not* carbolic acid, which precipitates albumen) to the
specimen to prevent decomposition.   Salkouski's fluid
is an excellent one for the purpose (" Deutsche Medi-
cinische Wochenschrift," xiii., No. 16, 1888).   It consists of a fluid containing 5 to 7·5 c.c. of chloroform in

a litre of water; about 2 oz. of this preparation should be used. A few drops of thymol or oil of turpentine also answer satisfactorily.

When the deposit has finally settled the supernatant fluid is carefully poured off. A pipette closed by the finger at the upper end is then introduced and some of the deposit allowed to enter by momentarily removing the finger. On the removal of the pipette the excess of fluid around the pipette is removed with a cloth, and the tube then allowed to rest for some seconds, standing perpendicularly on a glass slide; the heavier particles, such as crystals, will thus sink, and in this way a satis-factory sample of the deposit is obtained; a drop or two of fluid is then permitted to escape, and a cover-glass applied. The specimen should then be examined first with a half-inch (to detect the larger crystals) and then with a quarter-inch lens.

If the deposit be small, considerable assistance is afforded by adding a drop of a solution of iodine dis-solved in iodide of potassium (see p. 33) to the fluid on the glass slide, or a few drops (three are sufficient) of magenta solution to the deposit in the specimen glass.

The most convenient formula for the magenta is Woodhead's :—

Magenta . . . . . . 1 part
Rectified spirit . . . . 20 parts.
Distilled water . . . . 180 ,,

By either of the above means, the epithelial cells are brought more into prominence, and by their distribution in lines, or otherwise, may draw attention to other bodies, such as casts of the renal tubes, in which they are lying. In addition to this, should amyloid casts be present they will be stained a deep brown by the iodine solution.

M 2

Another advantage of using a stain is that the objects are more easy to focus when some colouring matter is employed. Unless this be done, the transparent cor-puscles, especially if the deposit is small, will often escape notice as the lens is gradually approached to the glass, with the result that the cover-glass is pushed hard on to the slide, and the fluid, carrying crystals, cells or other bodies with it, is forced from beneath the edges of the glass and that specimen rendered useless.

Directions were given above that the urine should be allowed to stand for 24 hours before the deposit is exa-mined.   In certain cases it is also advisable to look at a specimen about six hours after it is passed.   Some urines undergo decomposition very rapidly, or the urine may be drawn from the bladder actually decomposed. If " cayenne-pepper " grains be deposited these should also be examined as soon after they crystallise out as possible, in order to recognise the uric acid formation ; if the urine becomes alkaline from decomposition these crystals are transformed to those of urate of ammonia, and thus false impressions may be derived.

There is generally a slight deposit even with healthy urine, and if this be examined will be found to consist of epithelial débris, various crystals and leucocytes. Very often also there is a copious deposit of urates especially in the urine passed in the early morning.

The various objects met with in urinary deposits will now be described seriatim, the pathological significance of each being duly considered.

The sediments naturally divide themselves into **or-ganised** and **non-organised**; the former including epithelial cells, the corpuscles of blood, pus and mucus, renal casts, parasites and portions of new growth; whilst under the latter division will be described the

numerous forms of crystals which separate from the
secretion under different conditions.

ORGANISED DEPOSITS.

1. **Epithelium.**—As already stated some epithelial
cells are found in the very slight deposit which falls in
healthy urine ; these arise from the urethra and bladder,
and in females from the vagina, and it is only when
present in large numbers that their presence is of any
note ; but a different construction must be put on the
occurrence of renal cells.

FIG. 27.—Vaginal epithelium.

The cells for the most part are of the *squamous* variety,
are polygonal in shape and contain only one nucleus ;
distributed amongst them will be found their earlier
form, circular in appearance. These arise from the
meatus and prepuce. Those from the vagina are of
the same shape, but much larger (fig. 27).

Bizzozero describes a form of epithelium which has
for its origin the surface of the male urethra ; the cells
are cylindrical, have well-marked outlines, and gradually diminish in size towards their attached extremities,
and thus appear sometimes tailed, at other times
rounded, according to the surface which is presented to
the eye.

If any of the above cells are present in large numbers, some affection, most probably inflammatory, may be assumed, of the parts from which they come, thus, the cells described in the last paragraph are especially numerous in men suffering from an old gonorrhœa, and Sir William Roberts (*loc. cit.*, p. 122) points out that the deposits found in the urine of persons subject to nocturnal emissions have much the same appearance to the naked eye, a collection of whitish flakes and strings being found at the apex of the collecting glass.

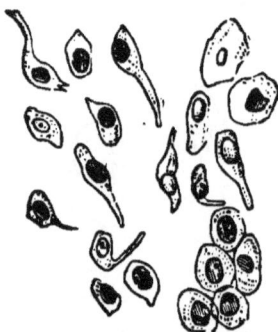

FIG. 28.—Epithelium from the bladder, ureter and pelvis of the kidney.

In women afflicted with leucorrhœa, large pavement epithelial cells are found in large numbers in the deposit, often united by their borders into patches of rude Mosaic.

The cells derived from the epithelium of the bladder, ureter and pelvis of the kidney (fig. 28) are difficult to distinguish from one another. A great variety of forms are found. Bizzozero and Eichorst maintain that the type of cells is the same for all these parts; the cells may be spindle-shaped, cylindrical, tailed, spheroidal oval or irregular in shape. They are smaller than those described above. Those that come from the superficial layers are spheroidal or oval; they possess one nucleus and are generally coarsely granular. The cells arising from the deeper layers are more irregular, or spindle-shaped, and often present processes projecting from their sides. The single nucleus is larger, and the protoplasm is more finely granular. In cases of vesical or renal calculi these cells are often found in large numbers; they also occur in the urine of patients suffering from pyelitis.

The recognition of *renal epithelium* (fig. 29) is of much more importance. The cells are of two chief

FIG. 29.—Renal epithelium, healthy and fatty.

varieties, according as to whether they are derived from the straight tubes or from the cortex, but the two kinds are often very difficult to distinguish from one another.

The typical cells from the cortical portion consist of a circular nucleus about the size of a red blood corpuscle surrounded by a finely granular protoplasm, the outline of which is generally indistinct. The cells from the straight tubes are flatter and more cubical in shape; the cell wall is generally plainly seen.

The cells, therefore, are rather smaller than those derived from the other parts of the genito-urinary tract. They are often aggregated together into small groups, or form complete casts of the renal tubules (see below). A few cells are frequently seen cohering to hyaline and other casts.

The presence of these cells indicate some lesion of the kidney substance, most probably inflammatory. If the cells show fatty degeneration, this is indicative of a similar process going on in the kidney, and thus is of grave import.

Prof. v. Jaksch ("Clinical Diagnosis," p. 179) describes another variety of cells which occurs in the convalescent stage of acute nephritis (of scarlatina and erysipelas). These are small round cells with an eccentric nucleus. He considers that they are doubtless the

Fig. 30.—Blood corpuscles in urine.

young kidney cells formed within the tubules in the process of repair.

2. **Red blood corpuscles** (fig. 30).—The occurrence of blood in the urine must always give rise to some anxiety, and a careful clinical examination should be made to try and ascertain its source. Much assistance may be gained by a diligent microscopical search.

The presence of blood may be determined by certain chemical tests, of which the most satisfactory is the guaiacum test. One or two drops of tincture of guaiacum are added to a small quantity of urine in a test-tube. and about half a drachm of ozonic ether is floated on the surface. If blood be present, a blue colouration

appears at the junction of the two fluids, which gradually spreads through the ether.

The most reliable test for blood in the urine is, however, the discovery of the red blood corpuscles by the
microscope. The corpuscles do not always retain their
form and colour, in fact they often appear much changed.
If the urine be of very low specific gravity, especially
if it become ammoniacal, the corpuscles gradually lose
their colour, swell by inbibition and then appear as
colourless rings and finally may disappear altogether.
If, however, the urine be of specific gravity 1020-1025,
they will retain their form for some days, or appear
smaller and of a deeper colour. Occasionally they
shrink and become crumpled and misshapen. They
occur apart from one another and do not run into
rouleaux.

The corpuscles, then, are characterised by their thin,
clear outline, the absence of a nucleus or cell contents
and their feeble refracting power.

As a rule there is not much difficulty in recognising
them, but occasionally other objects occur which by the
inexperienced may be easily mistaken for blood corpuscles.

The most likely bodies are the sporules of some fungi;
they may be distinguished by having a nucleus, and
also double forms caused by gemmation may be noticed.
These sporules also are not always circular, but are
often somewhat oval.

The nuclei of renal epithelial cells may easily be
taken for red blood cells; as already stated they are
of the same size and shape and not infrequently the
rest of the cells to which they belong cannot be made
out. This error may easily be rectified by the addition
of a drop of magenta solution (see p. 163) which will
stain the nuclei but not the corpuscles.

The most important point, having proved the presence of blood, is to attempt to ascertain its source. It may arise from the kidneys, renal pelvis, ureter, bladder or urethra.

As regards the causes no better classification can be found than that given by Sir William Roberts (*loc. cit.*, p. 141).

1. *Local lesions.*—External injury, violent exercise, calculous concretions, ulcers, abscesses, cancer, tubercle, parasites, active or passive congestion, Bright's disease.

2. *Symptomatic*, in purpura, scurvy, eruptive and continued fevers, intermittent fever, cholera, &c., mental emotion.

3. *Supplementary or vicarious*, to menstruation, hæmorrhoids and asthma.

That microscopic examination can be of great use in determining some of these causes is obvious, thus, small crystals of uric acid suggest calculous concretions ; numerous pus cells denote a possible abscess ; tubercle bacilli are conclusive proof of a tubercular process somewhere in the genito-urinary tract ; the discovery of parasites ; and finally the concomitance of renal casts, indicating Bright's disease. Some of these points will be again referred to.

Blood having its source in the *kidneys* produces certain characteristic appearances. The blood is intimately mixed with the urine, and generally does not form a deposit even after standing several hours, but occasionally a chocolate-coloured grumous sediment subsides. The secretion has a peculiar reddish or smoky appearance. The cells may sometimes be seen to have partially lost their colour, and present themselves as pale yellow rings, and the lesion then is probably an acute nephritis, either of fresh origin or an exacerbation of a chronic inflammation.

If the cells are few in number, and have quite lost their colour, a suspicion should arise either of miliary tuberculosis of the kidneys, or congestion of those organs.

In cases of abscess, embolism and tuberculosis, or from the presence of parasites, the hæmorrhage is usually slight, whilst if malignant disease is the cause, it may be very frequent and profuse. ·

Blood *from the pelvis of the kidney and ureter* has less marked characteristics, and very little help is afforded by the microscope. The occurrence of certain crystals, such as uric acid or oxalate of lime, may yield suspicion of a calculus, but the main evidence must be derived from the clinical aspects of the case. The same may be said if the blood comes from the bladder or urethra. When the bladder is the source, portions of villous growth may be found, or there may be other evidences of cystitis.

If the bleeding is from the urethra, it will probably occur independently of micturition. Under either of these circumstances the blood will appear of a bright red colour, will not be intimately mixed with the urine, and clots will very likely be found in the deposit ; if the clots are long and of about the diameter of a goose-quill, the ureter is the probable source, as such clots are occasionally formed there.

In the condition known as **" hæmoglobinuria,"** a chocolate-coloured sediment is deposited. No blood corpuscles are found in it, but it consists mainly of amorphous granular matter, which Sir William Roberts considers to be disintegrated corpuscles. Tube-casts are also present, mostly granular in appearance, accompanied by a few transparent fibrinous cylinders. Crystals of oxalate of lime occur, and the late Sir William Gull found large numbers of minute crystals of hæmatin.

4. **Leucocytes (pus cells)**, fig. 31.—A few isolated white blood cells will be found in any urinary deposit, but their presence only assumes importance when there is a sufficient number to justify the term " pus." The line to be drawn between the two conditions is naturally an arbitrary one, and no fixed rule can be stated. Generally speaking, no conclusions can be drawn from the occurrence of the leucocytes, unless

FIG. 31.—Pus cells in urine affected by acetic acid and unaltered.

there is a distinct deposit at the bottom of the collecting glass.

Urine which contains pus is turbid when first voided, but a sediment soon subsides. When the reaction of the urine is alkaline, as is usually the case, the deposit appears as a semi-gelatinous mass, which is not clearly separated from the general mass of the urine, and can be drawn out into long strings. Should the specimen be acid, the pus forms a uniform white layer, clearly defined, but easily miscible with the supernatant liquid.

The nature of the sediment is generally unmistakable, the only deposit likely to be confused with it is mucus. Pus may be distinguished from mucus by the addition of a little caustic soda, which converts the former into a viscid mass, which can be drawn out into long strings when the liquid is poured from one test-tube into another.

Under the microscope the pus cells retain their characteristic form—spherical in shape, and about one-third larger than a red blood cell. They are granular in appearance, and almost colourless, having only a slight yellowish tinge when crowded together. If the urine is very concentrated they appear small and crumpled. The addition of acetic acid causes the granules to disappear, and the nuclei to become prominent. They are often observed to contain fatty matter, which indicates that an abscess, of slow formation, has burst into the urinary tract from some neighbouring part, and not, as when occurring in epithelial cells, fatty degeneration of the kidneys.

If there be any doubt as to the nature of the cells, this may be cleared up by the addition of a drop or two of iodine in iodide of potassium (page 33) to a little of the deposit on a glass slide ; when a cover-glass is applied, and the specimen placed under the microscope, the pus cells will assume a deep mahogany-brown colour, whilst epithelial cells and detritus will only be stained a light brownish-yellow. The possible source of the pus is the same as in the case of blood, namely, the kidneys, pelvis of the kidney, ureter, bladder, urethra or genital tract, or from the rupture of an abscess into the genito-urinary tract from some other part.

Pus from the kidneys is rather rare, according to Sir William Roberts (*loc. cit.*, p. 499, *et seq.*), it may arise from (1) phlegmonoid inflammations ending in circumscribed abscess ; (2) multiple abscesses from purulent infection, and (3) occasionally from embolism, apart from pyæmia. The differential diagnosis of these conditions can only be arrived at from the clinical side.

With the aid of the microscope pus may be assumed t
come from the kidneys when the pus cells are accom
panied by renal epithelium, and more certainly when
they are accompanied by casts of the renal tubes.
Occasionally the cells themselves are moulded into this
form, and are known as "pus casts" (see below).

Pyelitis (suppuration in the pelvis of the kidney) is
only productive of a few cells in the urine, and the same
may be said when the pus comes from the ureter.   The
forms of epithelium accompanying the pus must be
carefully observed, and from these some conclusions
may be drawn.   With pyelitis the discharge of pus
may be intermittent, owing to some obstruction blocking
back the discharge.   A tumour, appearing and disap-
pearing, will then probably be made out in the loin.
An absence of mucus points to the mischief being
located above the bladder.   In the above conditions
the urine is acid.

The most common cause of pus in the urine is cystitis.
The local symptoms will most likely be quite sufficient
for the diagnosis without the aid of the microscope.
The urine is strongly alkaline, and often ammoniacal,
except in old cases, when it may retain its acidity.
The quantity of pus is often very considerable.   Mixed
with the pus cells, the numerous forms of bladder epithelium will be seen.

If the urethra be the source of the pus (as in gonorrhœa) the discharge will take place independently of
micturition, and a small quantity can generally be
squeezed from the orifice.   If pus be found in the urine
of a female patient, inquiries should be made as regards
the existence of leucorrhœa, before any further conclu
sions are drawn.

The bursting of a neighbouring abscess is indicated

by the sudden discharge of a considerable quantity of pus, and as before stated, under these circumstances the cells often contain fatty matter.

5. **Mucus.**—A translucent, flocculent cloud of mucus, separates from nearly all urines ; when the secretion is acid it remains thin and diffluent, but when alkaline, it becomes tenacious and ropy.

There is nothing characteristic to be observed microscopically, the mucous cells being indistinguishable from leucocytes. On account of its common occurrence, it has little clinical value; when present in large quantities it may indicate a catarrh of the urinary tract, the exact localisation of the trouble being indicated by the accompanying forms of epithelium.

In order to distinguish a deposit of mucus in acid urine, from one of pus, the caustic potash test should be applied. In the case of mucus, the deposit disappears, instead of being transformed into a stringy mass, as with pus. The mucin in which the corpuscles lie, is precipitated by acids and alcohol.

6. **Fat.**—Fat globules are but rarely found in urine. They may occur as accidental ingredients after the passage of a catheter. As a pathological curiosity they have been stated to appear after a too excessive or prolonged use of cod-liver oil.

Small globules of fat occasionally accompany casts, owing to the fatty degeneration which they have undergone, they also occur in the interior of epithelial cells (see page 168), and under similar circumstances, small globules may be found free in the deposit. In " chylous urine," large quantities of free fat are discharged. Globules of fat may then be distinguished under the microscope, but more commonly, only granular particles are visible.

In phosphorus poisoning also, free fat has been found
in the urine, which may easily be explained by the
intense degree of fatty degeneration which takes place
in this condition.

7. **Casts.**—If a specimen of urine is found to contain
albumen, it has become almost a matter of routine to
examine the deposit for "casts." This operation is a
most important one, for all casts (with the exception of
hyaline casts) are a sure indication of renal disease.
Although in nearly all cases casts accompany albuminuria, cases have been recorded in which these bodies
were found in non-albuminous urine. Thus, Nothnägel
("Deutscher Archiv für Klin. Med.," xii. 326, 1874)
has found them in cases of jaundice in which there was
no albuminuria, and Drs. Finlayson ("Brit. and For.
Med. Chir. Review," January, 1876), Burkart and
Fischl, have recorded similar cases. The form of these
bodies indicate that they have been moulded in the
renal tubules ; their size naturally varies greatly, according to the calibre of the tubule in which they
were formed, their diameter varying from $\frac{1}{1000}$ inch
(small) to $\frac{1}{500}$ inch (large), the largest number being
known as "medium sized," varying between these
limits.

Their nature is also very different, according to the
circumstances under which they arise. According to
their consistence they are known as blood, epithelial
and pus casts ; cylinders composed of micrococci ;
granular, waxy, fatty and hyaline casts, the last three
often being studded with crystals and epithelial cells ;
to these must be added cylinders formed of crystals and
urates, which may be known as "false casts," and
finally those strange bodies "cylindroids."

There seems to be some doubt as to the precise

structure of these cylinders. The sub-stratum of most of them is apparently an albuminous material exuded from the capillaries, whilst in some cases the casts are formed by degeneration and disintegration of epithelial cells and blood corpuscles, white and red. This seems to be the origin of the granular casts. Blood fibrin is probably the basis of the blood casts. Cornil has described hyaline globules in the epithelial cells, which are forced into the lumen of the tubules, and thus the hyaline variety may be formed.

The above objects will be described in the order named together with their clinical significance.

Blood casts (fig. 32a).—There can be no difficulty in recognising these bodies. The uniform size of the red blood corpuscles which usually retain their characteristic shape renders the casts conspicuous objects in the sediment, more especially as they are generally of the large variety. The discs may, however, become crumpled or otherwise deformed, which will render their nature somewhat obscure. They often appear thick and opaque owing to the presence of a large quantity of dark pigment. These casts are not usually perfect, but generally occur as fibrinous casts with the corpuscles scattered more or less irregularly over them.

The discovery of these bodies in a deposit has naturally much clinical importance. The most usual condition under which they appear is acute nephritis ; if sections of a kidney affected in this way are examined microscopically many of the tubes will be seen to be filled with blood and the formation of the casts can easily be imagined. Intense congestion, such as occurs after the administration of cantharides or turpentine will also produce them. Whether the congestion due to ob-

N

structive heart or lung disease will cause them to form is open to question.

These casts are occasionally seen after injuries to the kidney in which hæmaturia results.

Other rarer conditions accompanied by blood casts are :—renal calculus, infarct, new growth and tubercle.

Pus casts.—These are but rarely met with even when the amount of pus deposited is considerable.

They appear as cylinders composed of pus cells, closely packed together ; they are very rarely complete, and many cells will be found lying near them, having become detached.   It is most likely owing to this latter condition that they are so rarely found, as they very easily break up, and the majority of them are probably destroyed and the cells dispersed before the deposit settles.

The nature of the cells can easily be demonstrated by the use of reagents (see p. 173).

They are most constantly seen in connection with the discharge of multiple abscesses into the tubules ; consequently only very few occasions are presented for examining them.

Epithelial casts (fig. 32b).—When perfect, these cylinders consist of the characteristic cubical cells of the renal epithelium (see p. 167) closely pressed together. Care must be taken not to confuse them with blood casts, more especially as they occur under similar conditions ; careful focussing will exhibit the granular protoplasm surrounding the nuclei, and in cases of doubt a drop or two of magenta solution should be added to the deposit.   As with the blood casts, however, this perfect formation is not often seen.   More commonly the cells are scattered irregularly over a hyaline basis. The cells too, are generally broken or shapeless, or only the nucleus is visible. Immature forms of epithelium, oval or circular in shape are particularly common.

The presence of these casts indicate acute changes going on in the renal parenchyma, probably nephritis or congestion. This condition may either be primary or an acute attack upon old mischief; in the latter instance they will be accompanied by granular casts.

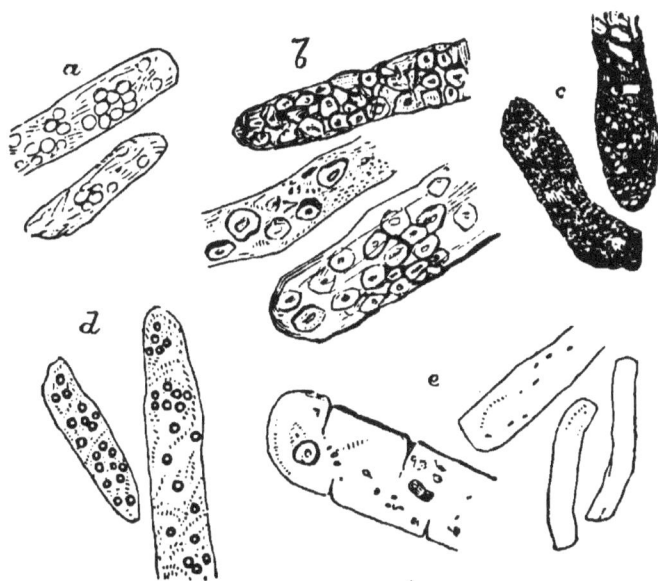

FIG. 32.—*a*. Blood casts. *b*. Epithelial casts. *c*. Granular casts.
d. Fatty casts. *e*. Hyaline casts.

As just hinted, cylinders composed of red blood cells and of epithelial cells are frequently found together.

Casts composed of micrococci.—These bodies are not often met with, and are very likely to be confused with finely granular casts. They are markedly opaque and have rather a greyish colour. Under a high power ($\frac{1}{12}$ immersion) they may be resolved into

N 2

their individual elements and the micro-organisms can
be distinctly made out. Apart from this they may be
distinguished by their resistance to caustic potash and
nitric acid.

They imply a very serious condition in the kidney;
most probably septic embolism; but they may also
arise from the extension upwards of septic pyelitis
(v. Jaksch).

Granular casts (fig. 32c).—These are more frequently met with than any other form. They are
opaque, usually have a sharp outline, and are grey or
brown in colour. Fragments only of these bodies are
usually found, and by their uneven extremities show
further signs of having been broken. There are two
distinct varieties; in one the granules are so fine as
only to be differentiated by a very high power, and in the
other the granules are much coarser, so that they can
easily be made out with a half-inch lens.

When present they indicate the later stages of an
acute renal affection or a more chronic process; and
may be said to be diagnostic of Bright's disease. When
the process is acute they are accompanied by epithelial
and perhaps blood casts; but when occurring alone,
contracted granular kidneys may be assumed. No difference can be drawn as to the anatomical conditions,
by observing whether the granules be fine or coarse.

Most authors now agree that they are formed by the
degeneration of the blood and epithelial casts.

Amyloid casts—These are clear, homogeneous
cylinders, highly refracting and possessing a clear
outline with rounded extremities. They are often of
a great length, but being brittle, are easily broken, so
that fragments are more commonly seen than perfect
casts. They are found of varying diameter, the largest

being almost an inch in breadth, whilst the smallest are not more than the breadth of a red corpuscle, and between these limits all sizes occur.

Owing to their transparency, these casts are difficult to see, and unless some staining solution, such as magenta or iodine in iodide of potassium be employed they are easily missed. Attention is sometimes drawn to them by the linear arrangement of crystals, leucocytes or epithelial cells clinging to them.

They occasionally exhibit the amyloid reaction (see p. 101) with iodine in iodide of potassium, or methylaniline violet, but this is by no means constant.

Their exact composition is not known but v. Jaksch concludes that they are very complex and may result from the destruction of epithelium, or from the exudation products (amyloid material, fibrin) into the renal tubules.

They sometimes occur in large numbers.

Their diagnostic significance is not great, as they are found in all forms of chronic renal disease; consequently amyloid disease of the kidney must not be diagnosed from their presence alone.

Fatty casts (fig. 32d).—Cylinders composed entirely of fatty globules are but rarely met with, although they do occur; more commonly the globules are found scattered more or less numerously over a hyaline matrix.

Fatty crystals frequently accompany them, being composed of crystals of the fatty acids or the combination of these with the earthy metals.

These casts are easily recognised by the sharply defined outline and high refracting power of the globules, but their nature may be proved by running a drop or two of ether under the cover-glass, when the globules will be immediately dissolved.

Fatty casts occur in connection with sub-acute and chronic nephritis, when the disease has lasted for some time. They indicate, especially when occurring as pure casts formed entirely of fat, an advanced degree of fatty degeneration of the kidneys, and consequently necessitate a grave prognosis.

Hyaline casts (fig. 32*e*).—These are very similar in appearance to the amyloid casts already described, but are not so highly refractive, and their extremities are somewhat more square. They are homogeneous, pale bodies, sometimes described as "ground-glass like," and are of varying length and thickness. They are even more difficult to see than the amyloid casts, and require to be stained with some medium, magenta being the best; a few drops of the stain should be added to the deposit (p. 163). Again, as with the amyloid cylinders, attention is often drawn to them by the arrangement of the epithelial cells, &c., adhering to them, and this is more marked after the use of the staining material. Their diagnostic value is small when occurring alone. Henle found them when the kidneys were quite healthy. Huppert discovered them in the urine of epileptic patients after an attack, no renal lesions being present.

When, however, epithelial or blood cells are found upon them, some conclusions as regards renal disease may be drawn. In such cases nephritis may be assumed, more or less acute. In fatty degeneration of the kidneys, fat cells are often seen thickly crowded together upon them.

Cylinders composed of urates, or crystals ("false casts.")—When examining a urinary deposit rich in amorphous urates, collections closely resembling granular casts are frequently observed; they are caused by the accidental accumulation of the urates, and are consequently not in any way connected with true casts. They may often be caused to form by gently moving the cover-glass once to and fro with the finger. Unless the observer is aware of their nature, they are not at all unlikely to be mistaken for casts; slight pressure on the cover-glass will immediately disturb the formation and cause them to disappear. Casts composed of hæmatoidin crystals have been described, but I have never seen them; they have been stated to occur in the urine of infants.

Cylindroids.—These bodies were first described by Thomas in 1870 ("Archiv. für Heilkunde," xi., 130) and are not uncommonly seen in the urine under various conditions. They are ribbon-like bodies, much longer than casts, and are often twisted upon themselves, and studded with small crystals and epithelial cells. In general appearance they are not unlike threads of cotton, but are easily distinguished from them by a more irregular outline, the absence of fraying at the ends, and the cells and crystals adhering to them in a way which they never do with the extraneous ingredients of urinary deposits.

Thomas first found these bodies in the urine of scarlet fever, but they often occur in Bright's disease, more particularly in the chronic forms; in the urine of children, too, they are not uncommonly to be seen, and Bizzozero has discovered them when there was no renal disease.

Their origin is doubtful, but they appear to be added to the urine after its escape from the kidney.

Casts of the seminal tubes.—These bodies are only seen in very rare instances. In general appearance they are similar to hyaline casts, but are much broader and usually have spermatozoa embedded in them

They are occasionally found in the deposit of the early morning urine of those patients subject to nocturnal emissions.

8. **Spermatozoa** (fig. 33).—Spermatic filaments when found in urine are seen as pear-shaped bodies possessing a minute oval head, not more than $\frac{1}{10000}$ inch in breadth, and a long tapering tail, slightly thickened at its attachment to the head, the entire

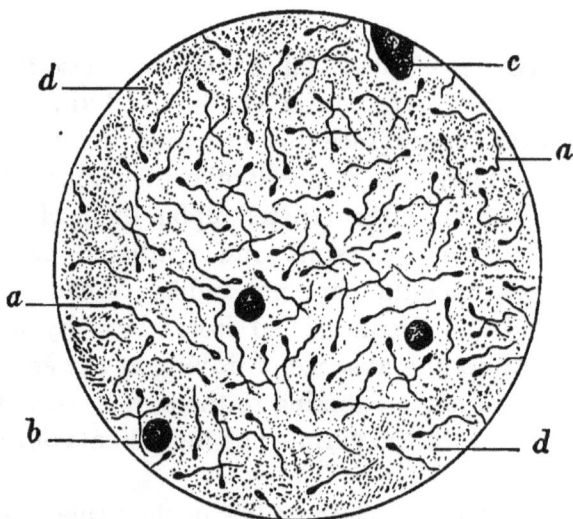

FIG. 33.—Spermatozoa (Ultzmann).

length being about $\frac{1}{800}$ inch. They are thus easily recognised by a quarter-inch lens. In urine they are always motionless, but when examined fresh they exhibit active movements.

In some instances semen escapes with the urine in such quantities as to form a mucous-like deposit, and whitish flakes are then often found, which may be removed with a pipette and the characteristic spermatozoa demonstrated. In medico-legal cases the presence of these bodies may be of the utmost importance, especially if found in the vaginal mucus.

In other respects they have very little diagnostic value, for many cases are on record in which they were constantly found in the urine for years, without any detriment to the general health of the patient. This is a matter of some importance, as quacks frequently impose upon their victims by exaggerating the danger of the occurrence of these bodies in the urine, and so reap a bountiful harvest whilst administering drugs for the ostensible purpose of averting the terrible fate which they predict for those who are unfortunate enough to consult them.

9. **Fragments of morbid growths.**—In the early days of the use of the microscope " cancer cells " were supposed to be so characteristic as to be easily recognised, and consequently great stress was laid upon their detection in cases of doubtful diagnosis. Now, however, the term is but rarely used, as the cells from a morbid growth differ so little from the ordinary transitional epithelial cells, that it is impossible to distinguish the former from the latter.

If a great number of large irregularly-shaped cells be found together with a blood-stained sediment, the diagnosis of a malignant growth may be suggested, but this is of no value unless the symptoms and physical signs also point towards the same conclusion.

In some rare cases small fragments of the growth may be discharged (especially in females) which are large enough to be submitted to a more complete microscopical examination, and then the existence of a growth may be determined. Those most likely to be met with are portions of villi from a superficial growth of the bladder, or fragments from the surface of a cancerous ulcer of the uterus,.rectum, or some other portion of the genito-urinary tract.

10. **Parasites.**—The parasites found in the urine naturally divide themselves into pathogenic and non-pathogenic. Every urine when allowed to stand will develop large quantities of bacteria, their form depending upon whether acid or ammoniacal fermentation is set up, but these do not necessarily interfere with the search for other more important organisms, such as tubercle bacilli and the ova of Bilharzia hæmatobia.

In collecting the urine intended for the search of micro-organisms, the parts about the orifice of the urethra should be carefully cleansed and the secretion be received into a perfectly clean vessel; the importance of these precautions is obvious. The deposit should be examined after the urine has stood for twelve hours. The parasites having a pathogenic value will be first considered, and then those which are non-pathogenic.

Pathogenic Micro-Organisms.

Tubercle bacilli.—The discovery of tubercle bacilli in the urine is of course of great importance, for it indicates that a tubercular process is proceeding either in the genito-urinary tract or its immediate neighbourhood.

After standing for several hours, the supernatant fluid is poured off from the deposit. A little of the deposit is removed by means of a pipette, and the outer side of the tube is wiped with a cloth; the open end is then allowed to rest for a few seconds on a cover-glass, one drop allowed to escape, and distributed over the glass, either by means of the pipette or more satisfactorily by placing another glass over the first, and gently sliding them apart. Owing to the dilution which the bacilli must undergo in the fluid, a considerable number of glasses should be prepared before a negative answer is given.

After the glasses are dry they are stained in precisely the same manner as will be detailed under the "examination of sputum." The various steps, however, should be carried out as rapidly, yet as gently, as possible, as the film does not become very firmly fixed to the glass, and the salts being easily dissolved may carry some of the other matters away with them.

If no bacilli be found, and yet the other features of the case point strongly to a suspicion of tubercle, the following method may be adopted :—

About 100 c.c. of the urine are mixed with 10 drops of caustic soda and boiled for five minutes. The mixture is then poured into a conical glass and allowed to stand for twenty-four hours. Portions of the sediment are then removed to cover-glasses and examined for tubercle bacilli in the ordinary way (Biechert).

If the result is still negative, it by no means follows that the disease from which the patient is suffering is not tubercular, but granted that bacilli are found, what conclusions may be drawn ?

In the first place, as already stated, it is quite certain that there is a tubercular lesion somewhere in the genito-urinary tract, or in some tissue immediately adjacent to it, which is indirectly affecting that tract. In every probability there is some ulceration, although bacilli have been found in miliary tuberculosis where no ulcers could be found.

The character of the predominating variety of epithelial cells will lend considerable aid in localising the lesion. If, together with renal epithelium, casts are associated, the kidney may be safely assumed to be the

188   MEDICAL MICROSCOPY.

organ affected, although it is very important to consider most carefully the history and clinical facts of the case, and take the microscopic examination as an *aid* to diagnosis, rather than an infallible test.

As a rule the bacilli are very few in number, but occasionally large colonies are found, but no deductions as to the severity and extent of the disease must be made from whether few or many rods are seen.

Erysipelas.—Various authors, notably, v. Jaksch and Fehleison (" Die Ætologie des Erysipels ") have described cases of erysipelas in which nephritis has supervened, and in which streptococci have been found in the urine when first passed, which are indistinguishable morphologically from the streptococcus erysipelatos. This condition ceased when the erysipelas disappeared.

Too much stress must not be laid upon this form of bacteriuria, as micrococci have been found in the urine in other septic diseases, such as ulcerative endocarditis and glanders so that they have no diagnostic value.

The spirillum of **recurrent fever** has occasionally been met with, but only when its association with blood pointed to the probability of hæmorrhage into the kidney and the consequent escape of the organisms from the capillaries.

The fungus of actinomycosis has also been stated to occur when the urinary organs are attached.

Septic micro-organisms.—These occur in all forms, rods and micrococci. They must be distinguished from those due to the decomposition of the urine *after* it has been passed.

In septic bacteriuria the urine is turbid when passed. The most frequent cause of this condition is the use of unclean catheters.

This subject has been most exhaustively studied by Sir W. Roberts, and the results of his observations are given in his work "On Urinary and Renal Diseases." He divides bacteriuria into four groups :—

Group I. *Bacteriuria associated with incipient putrefactive changes in the urine.*—The urine is feebly acid, neutral or feebly alkaline, and passes on quickly to decomposition. Rods are most frequently seen under the microscope. This condition is not infrequent in women suffering from leucorrhœa, in men who have suffered from stricture and who have frequently used catheters or bougies.

Group II. *Bacteriuria with ammoniacal fermentation of the urine.*—This indicates a serious condition. The urea is transformed into carbonate of ammonia, the fermentation being due to the action of the "micrococcus ureæ" (Cohn). The organisms occur singly or in chains of two, three or four cocci. This form is apt to arise in old stricture cases, in the presence of calculi, after operations, in cases of enlarged prostate and in all other conditions in which the bladder is unable to empty itself completely.

Group III. *Bacteriuria without decomposition of the urine.*— Rods and cocci are found, and the urine though cloudy on passing becomes clear on standing. The supernatant liquid continues transparent and acid for many days, and the organisms in the deposit show no signs of multiplying. All this leads to the inference that the seat of growth of the organisms is not the urine itself, but some portion of the surface of the urinary mucous membrane. The condition is associated with painful and frequent micturition and pains about the neck of the bladder.

Group IV. *Micrococci in the urine without decomposition.*—

This condition is extremely rare, Sir William Roberts having only met with one example.

Sarcinæ.—This seems to be the most suitable place in which to mention the occurrence of sarcinæ in the urine, although they cannot be strictly stated to be pathognomonic. The organism is discharged with the urine, sometimes forming a greyish-white deposit.

In appearance the groups of elements are similar to those found in vomit, but are somewhat smaller. The seat of their production is probably the bladder. The symptoms accompanying its growth are said to be lumbar pains and painful micturition.

Bilharzia hæmatobia (distoma hæmatobium).—The ova of this parasite are discharged with the urine ; the worm itself will be described in a later chapter.

The urine containing the ova is mixed with blood (endemic hæmaturia) and frequently contains much fat.

This disease only occurs in hot countries such as Egypt and the Cape, and those cases which have been watched in this country have contracted the disease whilst abroad. If the urinary deposit be examined, amongst the pus cells and débris will be found minute oval bodies, about $\frac{1}{170}$ inch long, furnished at the anterior end with a projecting spine, whilst the other is comparatively blunt.

In the mature ova the embryo may be seen as a granular ciliated body. Occasionally the free embryo may be seen either rapidly moving about, or slowly stretching the body out to its full length and then again retracting it. In such cases the empty shell of the ovum will be seen lying near the organism.

Filaria sanguinis hominis.—This parasite is more easily discovered in the blood than in the urine, and will be therefore described in the chapter on " The Examination of the Blood." In chylous urine this worm may, however, be found by examining the deposit in the usual way, and will be seen surrounded by blood and pus cells. As the animal is viviparous no ova are discharged.

Eustrongylus gigas.—This is the largest of the nematoid worms, but is so exceedingly rare in the human subject as to merit no more than passing notice. A specimen preserved in the museum of the Royal College of Surgeons of London is recorded as having been found in the human kidney after death. It is more common in animals.

In general appearance it much resembles the Ascarus lumbricoides, but is considerably larger, reddish in colour and possesses six oval papillæ instead of three.

Ascarus lumbricoides.—This worm is described in the chapter on the examination of fæces ; it is only found in the urine when a communication exists between the lower bowel and the renal tract.

Echinococci.—For a description of the microscopical characters of the hydatids and their cysts the reader is referred to the chapter on "The Examination of Discharges."

The occurrence of a hydatid cyst in the kidney is very rare, but the discovery of the hooklets, &c., in the urinary deposit may clear up an otherwise obscure diagnosis. The condition most likely to be mistaken for this disease is hydronephrosis.

Complete scolices are very seldom seen, as the vesicles are broken in the passage down the ureters, and only hooklets and portions of the laminated structure of the cysts can as a rule be made out.

The discharge of a cyst is usually attended with blood or pus in the urine.

With the use of the microscope alone the site of the disease cannot be ascertained, and a careful clinical examination must be made. The kidney is the most usual organ attacked but the cyst may be situated in or near some other part of the genito-urinary tract.

NON-PATHOGENIC MICRO-ORGANISMS.

When any urine is allowed to stand for some time it soon becomes crowded with micro-organisms.

Micrococci are usually most numerous, but rods are also seen (vibriones) frequently in active movement. The micrococcus ureæ is often found as almost a pure culture on the surface of the fluid.

Moulds are not so common, except on urines which contain sugar when they develop to a huge extent.

If the yeast fungus (Saccharomyces cerevisiæ) is found, it is a sure sign of sugar in the urine. This fungus occurs as oval budding cells, entangled in a mycelium of branching threads.

Another mould which is fairly common is ordinary mildew (Penicillium glaucum); this appears in acid urine which has been allowed to stand. It consists of oval cells and mycelia, but aërial hyphæ are soon sent off terminated by brush-like filaments arranged in groups, in which the spores are contained.

Infusoria.—Such organisms are never found in fresh urine, but only when fermentation has commenced. They are similar to those described in the chapter on the " Examination of Fæces."

UNORGANISED DEPOSITS.

Under this head will be described the numerous crystals and amorphous collections which occur in urinary deposits.

The majority of the substances are naturally dissolved in the urine, but are thrown out of solution either on account of being developed in excess, or by reason of a change in the reaction of the secretion or even of its temperature (urates). In addition there are a few additional constituents, such as leucin or tyrosin, which depend for their production upon some abnormal chemical process brought about by disease.

A conclusion as to the chief nature of the deposit may sometimes be drawn from its **naked-eye characters.**

Amorphous urates have a characteristic appearance. The sediment is loose and powdery, and varies in colour from a deep red to nearly white, according to the amount of urinary pigment carried down with it. A corroborative test for urates is the facility with which they dissolve with even slight heat; should any deposit remain after warming, the sediment was a mixed one; too much heat must not be applied, or albumen may be thrown down. Another deposit usually coloured is uric acid. The well-known "cayenne-pepper-like" grains are seen collected together at the bottom of the glass or adhering to its sides.

If the urine be alkaline, and phosphates be deposited, a brilliant white crystalline layer falls to the bottom of the glass, and crystals will also be noticed in wavy horizontal lines on the sides of the containing vessel. If the sediment be shaken up and some of the cloudy urine placed in a test-tube and boiled, heat tends rather

o

to increase the cloudiness than to diminish it, but the fluid immediately becomes clear upon the addition of a drop of nitric acid. The above remarks chiefly apply to a deposit of triple phosphates. If the urine be alkaline when passed, a peculiar pellicle of amorphous phosphate of lime not uncommonly collects on the surface of the liquid. When shaken, this layer breaks up into small plates, which exhibit a play of prismatic colours.

The only other deposit which has marked character-istics is oxalate of lime. This forms a double layer at the bottom of the specimen-glass, the upper being well-defined, dense, and white in colour, whilst the lower is more like a deposit of mucus, loose, and of a grey colour. The crystals also form lines on the sides of the vessel, as if the glass had been finely scratched.

In examining a deposit for crystals, the remarks made at the commencement of the chapter particularly apply. The outside of the pipette should be carefully wiped and the tube allowed to rest on the glass slide for a short time in order to allow the crystals time to fall to the point, and then the drop for examination may be allowed to escape.

These sediments are most easily classified as those occurring in acid urine, and those occurring when the secretion is alkaline.

SEDIMENTS OCCURRING IN ACID URINE.

Uric acid (lithic acid), fig. 34.—This is usually de-posited in crystals, but occasionally in a granular form. Crystals so minute as to appear as a flocculent cloud are occasionally seen after the addition of cold nitric acid to urine, and may be mistaken for a cloud of albumen.

The most common variety is the " cayenne-pepper-like '
grains already alluded to.

The primary form of uric acid is a rhombic prism or
lozenge, but this is so modified as to be hardly recognisable, and the number of varieties met with are
almost countless. If the angles of the crystals become
equal, quadrangular tables or cubes are produced.
Sometimes six-sided crystals are seen, much resembling
cystin, but differing from that substance in having two
of their sides larger than the others. If the angles are
rounded off ovoids or barrel shapes are obtained. At

FIG. 34.—Uric acid.

other times rods are seen, generally aggregated together
into stellate or fan-shaped masses. One ray is often
larger than the others, and forms a well-marked spike.
Other forms are seen too numerous to mention, but
whatever shape the crystals may assume they may
nearly always be recognised by their colour.

The natural colour of uric acid is white, but it generally carries down with it some of the urinary colouring
matter, and therefore the crystals are usually reddish-
yellow in colour, varying from a deep orange to a light
fawn according to the tint of the urine.

The urine is generally strongly acid, but if allowed to

stand until it is ammoniacal, the crystals seem to be gradually transformed into urate of ammonia, assuming the characteristic globular form; on the addition of a drop of hydrochloric acid they are re-dissolved, and uric acid again crystallises out. It is important, therefore to examine the original crystals soon after they are formed.

If there be any doubt as to the nature of the deposit the "murexide" test should be applied. A little of the sediment with a drop of nitric acid is evaporated to dryness on a glass slide over a spirit lamp. To the residue two or three drops of ammonia are added, if uric acid be present, a fine crimson colour makes its appearance, but sometimes not for two or three minutes.

The clinical significance of a deposit of uric acid is greatly dependent upon what time elapses between the passing of the urine and the appearance of the crystals In perfectly healthy conditons, uric acid separates out during the acid fermentation which occurs twelve or twenty-four hours after emission. If the sediment occurs within four or five hours, it indicates an excess but has still no special pathological importance. It is frequently thus met with in gout and chronic rheumatic complaints; in convalescence from febrile affec tions; in chronic diseases of the heart and lungs which interfere with oxidation of the tissues; in certain dis eases of the liver; in strumous and tubercular subjects in rickets, scurvy, and leucæmia.

In connection with gout, Sir A. Garrod has shown that the blood of a gouty patient contains an excess o uric acid, that it is deposited in combination with soda in the cartilaginous and fibrous tissues of the joints, and that in such a condition the kidneys themselves often suffer.

It is, however, when the uric acid crystals are deposited before the urine cools, or immediately after, that this sediment assumes great importance. It may then be feared that a similar process may take place in some part of the urinary tract, laying the foundation of calculi or gravel. Active treatment for the prevention of such a calamity is then immediately called for.

Amorphous urates (fig. 35).—This deposit was shortly referred to in the general remarks on unorganised deposits.

Urates are naturally colourless, but as deposited in urine they are nearly always coloured, the tint varying from orange or purplish to the lightest pink or fawn, but

FIG. 35.—Amorphous urates.

the sediment is always deeper in colour than the urine from which it separates. The deposit is loose, forming a uniform layer at the bottom of the glass, and is often accompanied by a peculiar film floating on the surface of the fluid.

The corroborative test is the application of moderate heat, which quickly causes the urates to dissolve. The chemical nature of the sediment is variable, but it usually consists of the mixed urates of potash, soda and lime (rarely of ammonia), the bases occurring in varying proportions. Under the microscope minute granules are seen, but these have no special characteristics.

The pathological significance of such a deposit is very

small ; its formation depends upon an acid state of the urine, and a certain degree of concentration. It is thus often observed in a state of perfect health, especially during the winter months and after profuse perspiration.

In disease it is very frequently met with, especially in febrile conditions, when it need give rise to no apprehension.

If urates are constantly deposited, without rise of temperature, there is probably organic disease somewhere in the body, either in the heart, lungs, liver or elsewhere.

FIG. 36.--Urate of soda.

Dyspeptic patients frequently complain that their urine is thick, especially that passed in the early morning.

Urate of soda (fig. 36).—This is a very rare deposit. The specimen from which the accompanying sketch was taken, was obtained from the urine of a patient suffering from gout. The sediment was yellowish in colour, and at first glance was mistaken for one of uric acid. Under the microscope the characteristic appearances were seen, consisting of circular opaque masses, from which spiny crystals projected. Some of the globes presented several such spines, others only one, variously curved, whilst still others were destitute of such projections. Some of the crystals seemed to be embedded in phosphatic plates. This sediment is chiefly found in the urine of children, and is deposited before the urine is passed. The spines, therefore, are very apt to cause irritation of the mucous membrane, and even to collect in masses and block the urethra, or form the nucleus of a calculus.

In rare cases, of which the above is an example, these crystals are found in the urine of gouty patients.

Oxalate of lime (fig. 37).—The general character of this deposit has already been given on page 194.

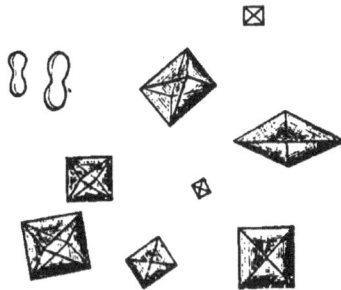

FIG. 37.—Oxalate of lime.

Under the microscope the crystals are seen to be transparent regular octahedra, or, especially in albuminous urines, they appear as dumb-bell-shaped bodies.

The octahedra are generally known as "envelope crystals," this shape being assumed owing to one axis of the crystal being shorter than the other two. They are seen of various sizes, but not unfrequently the crystals are exceedingly minute, and may easily be passed over without recognition. Occasionally they crystallise as pointed octahedra. The " dumb-bell " forms are in reality flattened, rounded discs, with a central depression on each surface, and according to

the way in which they lie in the field, appear as dumb-
bells, ovoids or circles, and may be made to apparently
change their shape by a gentle movement of the cover-
glass. These crystals are soluble in hydrochloric, but
insoluble in acetic acid.

Oxalate of lime occasionally falls as an amorphous
deposit, and can then only be recognised by its solubility or more elaborate chemical tests.

This sediment is frequently found in healthy urines,
especially after a vegetable diet, and consequently no
importance can be attached to its appearance, unless
it occurs persistently in large quantities, in spite of
alteration in diet.

The chief danger is its connection with the formation
of calculi. Much stress, however, has been laid on the
condition known as " oxaluria " and " oxalic diathesis,"
by Drs. Prout, Bird and Begbie, but Sir William Roberts,
Beneke, and others, have opposed these views, on the
grounds, firstly, that intense oxaluria may exist persistently without any of the symptoms attributed to the
oxalic diathesis, and secondly, that this group of symptoms may exist typically without the occurrence of
deposits of oxalate of lime in the urine.

Beneke further shows that oxalic acid has almost its
sole origin in the azotised constituents of the blood and
food ; everything, therefore, which retards the metamorphosis of these constituents occasions oxaluria.

Very little value, again, attaches to the amount of
deposit which occurs, for the urine may contain a large
proportion of oxalic acid, without it or its salts being
thrown out of solution. To ascertain the exact amount
present, a troublesome quantitative analysis would have
to be undertaken.

Sulphate of lime.—This substance is but rarely

found in urinary sediments. The author has never met with it. Fürbruger and Valentiner have described its occurrence in large needle-shaped crystals, and also in elongated tablets with abrupt extremities, they are sometimes arranged in sheaths and rosettes. Amongst them are seen masses of indeterminate crystalline structure. Sulphate of lime is also met with in urinary sediments in the amorphous form. It is insoluble in ammonia and acids, and its detection can only be assured by a rather involved chemical analysis, for which the reader is referred to textbooks on that subject.

The pathological conditions under which this deposit occurs have not been ascertained.

Phosphates.—The various forms of phosphatic deposit are met with in feebly acid urines, especially the " stellar phosphates " or neutral phosphate of lime, but as they are more frequently found in urines having an alkaline reaction, their description will be postponed until such sediments are considered.

Cystin (fig. 38).—Crystals of cystin are in rare cases found in urinary deposits. The urine from which it falls has generally a sweet-briar-like odour, and is turbid when passed. It is particularly liable to decompose, evolving sulphuretted hydrogen, which blackens white glass vessels, and the colour of the liquid sometimes changes from yellow to green. A great part of the cystin dissolved in the urine does not crystallise spontaneously, but may be brought down by the addition of a few drops of acetic acid.

To the naked eye the deposit is fawn-coloured, very much resembling urates. It is not dissolved by heat, and is insoluble in acetic acid, but soluble in ammonia (difference from uric acid). When the ammonia solution is allowed to evaporate, the cystin separates out as beautiful six-sided crystals, taking the form either of prisms or plates, the former being single or arranged in stellate groups.

To corroborate the presence of cystin in the deposit, the urine should be filtered, the sediment on the filter paper washed with a little water, and then transferred to a piece of platinum foil and heated in the flame of a Bunsen burner. Cystin burns with a bluish-green flame, emitting white acrid fumes, having an offensive odour resembling garlic, but does not melt.

A spontaneous deposit of cystin, when placed under the microscope, exhibits six-sided plates, showing lines

Fig. 38.—Cystin.

of secondary crystallisation; they are not often seen singly, but are placed on one another, or joined together in long strings. They closely resemble one form of uric acid, but differ in having all their sides equal.

Its occurrence in the urine is chiefly of importance as indicating the possible formation of gravel or a calculus. It has a peculiar tendency to run in families, especially affecting children and young male adults. Cystin may persist in the urine for many years, perhaps only appearing at intervals, without any injury to the general health, but merely causing physical irritation to the bladder and urethra by the formation and passage of small concretions.

Xanthin.—Bence Jones, Douglas Maclagan and Budd, have described this substance in urinary deposits. Bence Jones discovered it in the urine of a lad who three years previously had exhibited the symptoms of renal colic. The crystals resemble one form of uric acid, being pointed ovals, very much like whetstones. They are soluble in ammonia and insoluble in acetic acid (difference from uric acid). When the deposit is heated with nitric acid, a yellow mass is deposited on evaporation, which turns a violet-red colour when caustic potash is added. The clinical significance is not known, beyond indicating the possible formation of a calculus composed of this substance, of which there are now several on record.

Tyrosin (fig. 39).—Together with leucin, the crys-

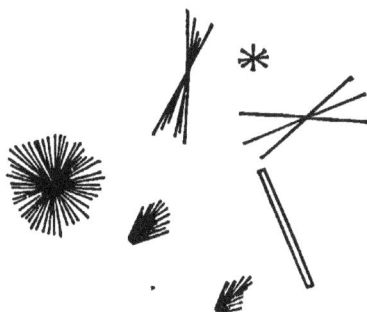

FIG. 39.—Tyrosin.

tals of which will be immediately described, this substance has been found in urinary deposits.

It occurs as sheaves of very fine needles, sometimes in yellowish-green crystalline globules.

If the nature of the sediment is doubtful, it should be dissolved in hot ammonia, and the following tests applied :—

1. If a little of the solution be allowed to evaporate,

tyrosin will crystallise out in fan-like groups of colourless needles.

2. Some of these crystals are dissolved in hot water, and to the warm solution some nitrate of potash and mercuric nitrate are added. A red precipitate, together with a red colouration of the fluid, indicates the presence of tyrosin.

3. Some more of the crystals are placed in a small capsule and a drop or two of sulphuric acid added. After standing for half-an-hour water is added. The solution is now boiled and saturated with carbonate of lime. The mixture is filtered and to the filtrate a little perchloride of iron solution (which must be free from acid) is added. In the presence of tyrosin a deep purple tint is produced.

A spontaneous deposit of tyrosin is rare, but it may be obtained from urine which contains it by the addition of acetate of lead, until a precipitate is no longer produced, and then passing sulphuretted hydrogen through the mixture. The precipitate is separated by filtration, and the clear solution which passes through concentrated by evaporation, when the tyrosin will crystallise out.

This substance is principally found in the urine of patients suffering from acute yellow atrophy of the liver, but has also been met with in phosphorus poisoning, typhus and other infectious fevers, also in leucocythæmia and certain nervous states.

Leucin (fig. 40).—As an urinary deposit this substance is still more rare than tyrosin, although it usually occurs in solution with that substance in the above named conditions.

The sediment, when it occurs, consists of yellowish oily looking drops, which have a concentric appearance.

The urine should be evaporated to dryness, and the residue heated with boiling alcohol. On cooling, the solution deposits shining white delicate plates of leucin, which are greasy to the touch ; they are insoluble in ether and chloroform (difference from cholesterin).

As a confirmatory test, some of the crystals may be dissolved in boiling water, and the solution heated with proto-nitrate of mercury, when, if leucin be present, metallic mercury will be deposited.

Hippuric acid.—Crystals of hippuric acid appear as colourless, four-sided, rhombic prisms or needles, the

FIG. 40.—Leucin.

former forming when it separates from a dilute, and the latter when from a strong solution. They are soluble in alcohol and insoluble in acetic acid, thus differing from uric acid and phosphates respectively.

They have no clinical significance. They occur in large numbers after a patient has been taking benzoic acids or after eating certain fruits.

Hæmatoidin and bilirubin.—These substances are both found in urine, but it is extremely difficult, if not impossible, to distinguish them one from another, their crystals have the same form, and no accurate chemical tests have been discovered which one will re-act to and the other not. The crystals are of two kinds, clusters of needles or minute rhombic tablets. Their colour varies from a light yellow to a deep ruby-red.

They are soluble in caustic soda. On treating the crystals with a little nitric acid, a green rim forms round them, and their red colour gradually changes to blue (Hoppe-Seyler).

These two substances occasionally occur in the amorphous form.

Hæmatoidin crystals have been found in severe acute nephritis, in carcinoma of the liver, in some of the acute specifics, such as scarlet fever and typhoid, and in jaundice. If they occur in large numbers, a previous hæmorrhage is indicated, or an abscess may have burst into the urinary passages.

Bilirubin has chiefly been discovered in the urine of jaundice.

Soaps of lime and magnesia.—In the urine of various diseases, v. Jaksch (*loc. cit.*, p. 201) has found crystals similar in form to those of tyrosin, forming rosettes of fine needles. Their nature is uncertain, but this observer considers that they are probably composed of the lime and magnesia salts of the higher fatty acids. They usually occur in the urine of febrile conditions, such as severe puerperal septicæmia.

SEDIMENTS OCCURRING IN ALKALINE URINE.

Triple phosphates (fig. 41).—The most typical form in which the phosphate of ammonia and magnesia ($MgNH_4PO_4 + 6Aq.$) occurs in urinary sediments, is a triangular prism with bevelled ends; but this is often considerably modified by the removal of the angles or the hollowing-out of the sides. Some of the crystals are more quadrilateral in form, whilst others appear almost like octahedra in consequence of the central part of the crystals not being developed. If the urine be highly ammoniacal, the crystals may have a stellate arrangement, consisting of feathery rays, like snow-flakes, or may form peculiar jagged figures like leaves.

Like all phosphatic deposits these crystals dissolve readily in acetic acid.

They cannot be said to possess any pathological value, their presence simply indicating an excess of alkalinity of the urine.

Neutral phosphate of lime (stellar phosphates), fig. 42.—The chemical formula of this substance is $CaHPO_4+2Aq.$ The crystals are formed of rods or

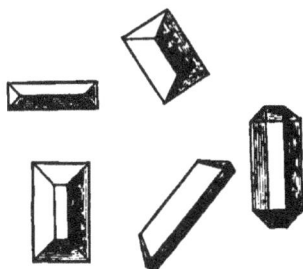

FIG. 41.—Triple phosphates.

needles varying in size, often showing lines of secondary crystallisation, and arranged singly or in sheaths and rosettes. They are generally surrounded by small crystals of oxalate of lime. The urine is, as a rule, feebly acid, or just becoming alkaline. The deposit is decomposed by ammonia and dissolved by acetic acid. This sediment is rarer than the other forms of phosphates and has a rather more grave significance. Sir William Roberts states that according to his experience "the presence of this deposit in quantity is an accompaniment of some grave disorder." The author has twice met with it in cases of advanced phthisis.

In scanty numbers the crystals may be met with in healthy urine, especially during the "alkaline tide" after meals, and if the secretion be rich in lime.

Basic magnesium phosphate.—These are likewise but seldom met with, chiefly owing to its easy solubility. The crystals occur as strongly refracting plates, having the form of elongated rhomboidal tablets.

Nothing is known as regards their pathological value, but they simply seem to indicate that the urine is concentrated.

Amorphous phosphate of lime.—This is the most common deposit when the urine is alkaline from

FIG. 42.—Stellar phosphates.

fixed alkali; crystals of the triple phosphate often accompany it.

In general appearance the sediment is not unlike the ordinary urate deposit, except that it is not coloured by the urinary pigments, and is consequently whiter than the supernatant liquid.

Under the microscope minute granules are seen, resembling the amorphous urates, collected together into irregular groups.

Such a deposit is frequently seen in healthy urine after a meal, and is a sure sign that the secretion is alkaline, and thus is a ready test of the success of drugs which have been administered in order to secure an alkaline reaction of that fluid. The urine is turbid when first passed, and the deposit rapidly subsides, an iridescent film collecting on the surface. Amorphous phosphate of lime (bone-earth) forms a large portion of phosphatic calculi, but requires some nucleus around which to accumulate.

Carbonate of lime.—This substance is always deposited in the amorphous form from ammoniacal urine. The granules are often collected together into dumb-bell-shaped masses or small concretions. The sediment dissolves in acetic acid with the evolution of gas, thus distinguishing it from the amorphous phosphatic deposit which dissolves in acetic acid without the evolution of gas.

When voided as gravel or small calculi, carbonate of lime is found in the crystalline form as small spheres exhibiting a radiating appearance.

Urate of ammonia.—Urate of ammonia is thrown down when the urine becomes strongly ammoniacal. The deposit may easily be obtained by allowing urine which contains the amorphous urates or uric acid to stand until ammoniacal fermentation commences.

To the naked eye the deposit is white, but occasionally it possesses a beautiful violet hue (Roberts).

Under the microscope dark, opaque, globular masses of various sizes are usually seen. More rarely this substance takes the form of minute dumb-bells, arranged singly or in groups.

If the deposit be dissolved by the addition of a drop of acetic acid, and a few drops of the solution placed on a glass slide, rosettes of uric acid will crystallise out, the crystals being quite colourless.

Cholesterin (fig. 43).—The occurrence of cholesterin in urine was first pointed out by Dr. Lionel

P

Beale ("Microscope in Medicine," p. 355). This author showed that the oily particles so frequently seen in the deposits from the urine in chronic Bright's disease contain cholesterin, which may readily be separated in a crystalline form.

But as a spontaneous deposit, cholesterin is extremely rare. It has been found associated with pyuria. The crystals take the form of rectangular plates with a notch in one corner, and commonly appear superimposed one on another.

Indigo.—This substance occurs either as the colouring matter of deposits, such as urates, or pure, in the

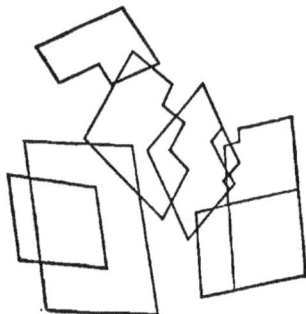

FIG. 43.—Cholesterin.

amorphous or crystalline form. The crystals are usually needle-shaped and arranged in rosettes. It not infrequently occurs in putrescent urine as blue shreds and films on the surface of the liquid or adhering to the sides of the glass. The presence of indigo is due to the decomposition of indoxyl.

It has no decided clinical significance.

Gravel and calculi.—The composition of these bodies is more easily determined by chemical analysis than by the use of the microscope.

Reference has already been made several times, whilst

describing the various deposits to the possible formation of calculi. Thus gravel is most usually composed of uric acid or of urates, or of the two combined, more rarely of cystin or xanthin.

Calculi may be formed of the mixed phosphates, carbonate of lime, phosphate of lime, uric acid, urates (chiefly found in the kidneys of children), oxalate of lime ("mulberry calculi"), cystin or xanthin. Various other concretions are occasionally met with formed of fibrinous material, blood, fatty matter or cholesterin.

As regards the microscopical evidence which may be obtained from scrapings of these bodies, reference must be made to the different heads under which the crystals are described. But as such scrapings will, in most cases, merely yield an amorphous powder, but little aid is rendered in the recognition of the stones, and it is therefore advisable to proceed at once to a chemical analysis.

Extraneous matters.—Matters entirely foreign to the urine, which occasionally find their way into the urine, are of two classes, the first of which is of great importance, whilst the second is purely accidental.

In the first class are included such substances as are the result of disease, or of a morbid process in some adjacent organ. Thus, substances belonging to the fæces may be found under such circumstances that they could not have been passed by the bowel, thus indicating a communication between some point of the urinary tract (most probably the bladder) and the intestine.

In other cases hairs have been passed, which, by the malignant course of the case, have been shown to come from a dermoid cyst. Fragments of tumours also occur, which have their source, not in the urinary organs, but in some neighbouring tissue.

The second class of extraneous matters includes any substances which have accidentally found their way into the urine, or in some hysterical cases, have been purposely placed there for the purposes of deception.

It is always advisable, therefore, tó cover specimens which have been set aside for examination, for otherwise, in hospitals especially, many small particles derived from the bedding, floors, food, &c., may fall in and be a source of much perplexity, before their true nature is ascertained.

One of the most puzzling of these objects is caused by the use of dusting powders, especially in female patients.   Starch grains are then seen, more or less altered by inbibition of fluid, and may be rather difficult to recognise ; if a drop of iodine solution is added, however, their change in colour to a deep purple will at once explain their nature.

Fragments of pine wood from the floor, becoming soft and swollen by soaking, often give the appearance of casts, unless carefully examined.

Other substances ·encountered, and with which the observer should make himself acquainted, are hairs, both human and from animals (especially cats and dogs), fibres of cotton, worsted, silk, wool, &c., these being often coloured ; portions of feathers ; bits of tea-leaves, showing cells and spiral vessels ; particles of food, meat fibres, oil globules, &c.   Young pediculi, and the larva of the blowfly (these last especially in diabetic urine) are also sometimes found, and hysterical people will add flour, sand and other powders, milk, or blood from the finger, in order to endeavour to deceive the physician and excite sympathy, generally of a practical form, for some pretended malady.

Preservation of urinary deposits.—A question

often asked is: " What is the best mode of making permanent preparations of urinary deposits "? This applies more especially to the various forms of crystals.
The ordinary mode of mounting with Canada balsam is
not satisfactory, this medium being too transparent for
the purpose.

The best plan is to use some preservative fluid in
which the crystals are not soluble. For uric acid,
oxalate of lime, urates of soda and ammonia, as well
as for casts, glycerine, diluted with water, answers as
well as any, or a dilute solution (1 in 50) of carbolic
acid may be used.

The method introduced by Dr. Lionel Beale is an
excellent one and is carried out as follows (" Microscope
in Medicine," p. 371) :—

A shallow " cell " has to be prepared. The most
convenient form for the purpose is that which is made
by painting upon a glass slide, with a fine brush, a
narrow border of Brunswick black, enclosing either a
square or circular space as may be most convenient.

The urinary sediment is allowed to subside to the
bottom of a conical glass, the supernatant liquid being
poured off, and a small quantity of the preservative
solution added. The deposit is again allowed to subside, and the solution poured off and replaced by a fresh
quantity. This method is adopted in order that the
deposit may become thoroughly saturated with the preservative fluid, otherwise there is danger of the preparation deteriorating before many months have passed.
Glycerine, if used, must be added to the deposit in the
conical glass in very small quantities at a time, so that
cells, casts, and other small bodies, may swell out again
after they have been caused to shrink.

After the final subsidence of the deposit, a small

portion may be removed with a pipette, placed in one of the cells above described, and the glass cover applied to the surface of the liquid, care being taken that the whole surface of the glass is well wetted with the solution, in order that no air bubbles may be included in the preparation. Any excess of fluid is now to be soaked up with a clean cloth or filter-paper, and the cover cemented to the cell by applying a little Brunswick black or Dammar varnish with a camel's hair brush.

Crystals of the triple phosphate or cystin have to be preserved in a different manner. The cell is made as before. The triple phosphate may be kept in water, to which a little ammonia and chloride of ammonium have been added. In this solution the crystals preserve their beautifully smooth character, while in pure water the surface becomes roughened.

Crystals of cystin are best preserved in organic acids, a very weak solution of acetic acid keeping them quite unchanged.

CHAPTER XV.

Examination of the Fæces.

UNDER the term "fæces" will be understood all those substances passed by the bowel, and consisting of the residue from the processes of digestion, mixed with the products of the secretions of the alimentary canal.

In this chapter will be considered, not only the matters which are expelled as the result of alterations in the dejecta produced by disease, but also objects such as parasites, which are added to or passed independently of the stools.

Before passing to the microscopical examination, a short epitome will be given of the chief alterations in the macroscopical characters of the fæces.

The character of the stools in health is greatly dependent upon the nature of the food which is taken, as will be seen by the following remarks.

No definite conclusions can be drawn from the **quantity.** The average in health is about five ounces daily, any considerable increase being principally due to the addition of fluid, which, however, bears no relation to the quantity taken by the mouth during the day. Naturally, the quantity varies considerably with the solid food that has been eaten.

The **colour** changes more than the other characteristics. In this respect the nature of the food has a marked influence. The usual brown colour be-

comes lighter with milk diet, and darker when much
meat has been taken; and the same effect is produced
when the fæces have been retained for some time in the
intestines.   An excess of green vegetables also communicates their colour to the dejecta.   Cocoa-nibs or
chocolate often cause the stools to assume a greyish
tinge.   Coffee, claret and porter, especially when these
liquids are taken in excess give their particular hues to
the contents of the intestine.

The colour is further considerably altered when
various drugs have been administered.   Iron, mercury,
manganese and bismuth cause a black colouration
owing to the formation of the sulphides of the metals.
Calomel yields a green, rhubarb, santonin or senna a
yellow, and logwood a red colour.

Various diseases produce such changes in colour as
to almost render them characteristic.   The stools are
almost colourless in obstructive jaundice, when they
have a pale greyish, putty-like tint.   In obstruction of
the pancreatic duct, too, the stools are almost white :
this being due to the fact that the action of the pancreatic secretion is essential to the production of the
ordinary fæcal colouring matter (Walker, "Trans. Roy.
Med. and Chirurg. Soc.," vol. lxxii., 1890, p. 257).

In the disease known as psilosis or sprue, the motions are pure white, with occasionally a layer of
yellowish matter over them; they are partially or wholly
formed according to the stage of the disease.   The
"rice-water" stools of cholera are also almost colourless,
but are more characterised by the consistence.   In
typhoid fever the motions are the colour of pea-soup.
The occurrence of blood in the evacuations, on the
other hand, produces a black colour (melæna), if it is
retained for some time in the bowel, but if due to the

bursting of a vessel through ulceration it retains its red colour.

In dysentery the stools vary considerably with the different stages of the disease, being sometimes merely streaked with blood, at others consisting of almost pure blood, whilst sometimes they resemble meat-washings.

Bile pigment not unfrequently makes its appearance, and is always pathological, depending upon some change in the liver or gall-bladder.

Green stools are principally seen in intestinal affections of children. As already stated this may be owing to calomel having been administered, but apart from this they may occur under two other conditions. Firstly, in infants diarrhœa is easily caused by excess of acidity, and the green colour which is then observed is caused by the presence of a bacillus (Lesage), which can be cultivated on artificial media, and causes blood poisoning when injected into animals. Secondly, the condition of green stools in children is sometimes brought about by the presence of biliverdin.

The **consistence** of the fæces depends partly upon the food which has been taken, but chiefly upon the quantity and nature of the secretions of the large intestine.

The **reaction** is normally alkaline, but an acid reaction does not necessarily imply disease. In young infants errors in feeding frequently produce acid stools, with diarrhœa and excoriation of the buttocks, whilst the " clayey " stools are usually alkaline, owing to the presence of carbonate of ammonia, the result of decomposition.

The **odour** of the motions yields little information. In children, a fœtid or acid smell often indicates that they are not being fed properly. The absence of bile, by withdrawing its deodorising action, causes a very bad odour.   In sprue, the smell is almost characteristic being particularly pungent and penetrating.

In stools of dysentery and melæna, and in any condition in which decomposition sets in rapidly, the odour becomes excessively fœtid.

Passing next to objects visible to the naked eye and not normal constituents of the fæces, we may mention first **foreign bodies.**   These are of all kinds, and the case-books of lunatic asylums record many extraordinary things which have been swallowed and afterwards passed by the bowel.

Among the most common of such bodies are portions of indigestible food, as seeds and stones of fruits, false teeth, masses of hair, small stones and portions of metal, &c.   When there is overaction of the small intestine (lientery) undigested food is nearly always found in considerable quantities.

In order to separate these and the following objects the fæces should be placed in muslin over which a stream of water should be directed; by these means most of the solid matter will be broken up and the larger bodies easily detected and separated.

Mucous and fibrinous shreds. — Nothnagel (" Beiträge zur Physiologie und Pathologie des Darmes," p. 185) has described an affection of the bowel which he has termed "enteritis mucosa or membranacea." In this disease cylindrical shreds of mucous membrane, sometimes forming complete casts, are passed by the rectum, accompanied with violent straining. After retention in the lower part of the bowel, they may be consolidated into " hard, white, rounded masses, about the size of nutmegs " (Fagge).   Dr. Lionel Beale (" Microscope in Medicine," p. 291) has described

several such cases.   Some of the shreds seemed to consist of firm mucus, in which the epithelial cells from the
large bowel and mucus corpuscles were embedded.   In
some of the cases the substance passed was firm enough
to form a tough leathery membrane, and under the
microscope exhibited fibres, mucus corpuscles and im-
perfectly formed epithelial particles.

Occasionally the casts are more distinctly fibrinous and
are tougher and ribbon-like; their form and character
can be demonstrated by removing them to a large dish
of water and gently moving them to and fro with a
needle, when they will float out and exhibit their whole
structure.   On careful examination pits may be made
out corresponding to the mouths of glands.

**Portions of bowel separated from intus-
susceptions.**—These are more interesting from a
purely pathological than from a clinical point of view,
as the diagnosis of intussusception will in all probability
have been made before the cast off portions are passed
by the rectum.   When floated in water they will be
seen to involve the complete circumference of the intestine; they are generally gangrenous and emit a most
offensive odour.   Under the microscope the more or
less altered structure of the part of the bowel from
which they are derived may be demonstrated; the
examination will be made more easy if a drop or two
of methylene blue be added to the water in which the
specimens are teased out.

Intestinal concretions.—These objects are usually the result of accident, and are mere pathological
curiosities possessing no clinical value.   Hard masses
are often passed with the motions by lunatics; thus Dr.
Langdon Down records a case in which such a mass
when examined proved to be cocoa-nut fibre.   In other

cases oat-hairs have been found matted together into a firm solid ball. In rare instances, drugs which have been administered for too long a period, or incautiously, produce similar results, and masses of carbonate of magnesia or insoluble salts of iron may be found.

Mention may here be made of the occurrence of "**scybala**" in the rectum, which often give much trouble, especially in elderly people. These are caused by the retention of fæces and removal of moisture, and the combination of the constituents of the excrement with inspissated mucus.

Gall-stones.—The passage of gall-stones is of great clinical interest, inasmuch as they often clear up a doubtful diagnosis. They are easily detected with the naked eye, and for closer examination may be obtained by shaking up the fæces with water, when they will immediately sink and are thus easily removed.

Stones consisting of bile-pigment are the most common. They often occur in large numbers. They are black in colour, irregular in shape, and sometimes present facets where they have rubbed against each other. They are friable and are easily crumbled. They vary much in size, but are usually about the size of hemp seeds. They are generally deposited round a nucleus, which consists of inspissated mucus.

Gall-stones are more rarely light in colour, and then most probably composed of cholesterin. Their structure can easily be exhibited by scraping off a little with a knife and adding a drop of water. The characteristic crystals (see fig. 43) are then seen, consisting of rectangular plates, with a notch out of one corner, superimposed one on another. Their nature may be further corroborated by adding to some of the scrapings a drop of sulphuric acid, when a brilliant red colour will make its appearance. These calculi are smooth, and some-times attain such a large size as to occupy the entire gall bladder. They are then of course only discovered after death. When several are present, they are facetted, and are much harder than those formed of bile-pigment. More rarely bile-stones are composed of carbonate or phosphate of lime.

Before concluding the account of the macroscopic examination of fæces, mention must be made of the bodies described by Virchow and Nothnagel (*loc. cit.*, p. 96) resembling cooked sago grains or frog-spawn. Virchow thought that they came from an excess of starchy food.

Finally the bodies known as "sable intestinal," and described by Dr. Sheridan Délapine in the "Pathological Transactions," xii., 1890, p. 111, deserve brief notice.

Labouline, in 1873, described small irregular granules, not unlike cork-dust, which he had found in the stools, and gave to them the name "sable intestinal." Dr. Délapine has investigated several cases of this kind.

The granules or small masses vary considerably in size, some being very minute, whilst others measure ⅜ inch in their largest diameter. In one case they were brown in colour, and were shown to be composed of vegetable cells, such as are found in sclerenchymatous tissue, a small amount of carbonate of lime, phosphates and oxalates, and a few silicious particles. In another instance they resembled coarse bran in appearance, and in yet another, small gall-stones. In the last case, however, it was shown that their nuclei consisted of undigested food matter, and that therefore they must have been produced in the intestine, and also that they were not composed of cholesterin, or of any of the products which may accumulate within the gall bladder.

After careful investigation Dr. Delapine came to the conclusion that all these masses were practically nothing else than residues of articles of food (fruit pips, &c.) which had not been digested, either because they were undigestible, or because the stomach and intestine failed to digest them. Yet they could not be said to be simply undigested matters such as those found regularly in the fæces, since by their accumulation they gave rise to suspicious appearances and symptoms.

Microscopical Examination.

The stools should be examined in two ways. First, by placing a small particle on a glass-slide, covering it with a cover-slip and distributing it into a uniform layer by gentle pressure. Secondly, by shaking up a portion with many times its bulk of water, allowing the sediment to settle, then removing portions by means of a pipette as in the examination of urinary deposits.

In the **stools in health** various particles of undi-gested food, animal and vegetable, are met with, and the student should make himself well acquainted with these, although it is true that mistakes are not so liable to occur in the examination of the fæces as in that of the sputum or urine. They are much the same as those which will be described as occurring in vomit *(q. v.)*.

When the individual from whom the specimen is obtained is on a meat diet, **muscular fibres** will invariably be found, but are sometimes so altered as to be hardly recognisable, being swollen and stained yellow by the bile; with a little care, however, their characteristic striation can usually be made out. With these fragments **elastic fibres** will also be seen ; they may be recognised by their clear outline, tendency to branch dichotomously, and the bold curves which are nearly always apparent. They are straighter and coarser than those described as occurring in human sputum, are very elastic, and are usually seen in masses.

Bundles of **areolar tissue** will also be observed after meat has been taken, especially if the digestion is at all faulty; the fibres are irregularly arranged, and generally appear in a shapeless mass the exact structure of which is difficult to make out. **Fat** is usually found in the form of crystals, the needles being arranged in rosettes; these are especially numerous when the pancreas is diseased; indeed their presence in combination with almost colourless stools is strong corroborative evidence of blocking of the pancreatic duct, but the exact nature of the lesion can, of course, only be made out clinically.

Fat also occurs in globules. In chronic alcoholism it is found in abundance in the stools.

When vegetables have been mixed with the diet, **vegetable cells** of various kinds are seen, spirals, hexagonal cells, single or in groups, &c.; the forms being too numerous to describe and depending upon the particular vegetable which has been eaten.

Starch granules.—Owing to inbibition these bodies are at first sight difficult to recognise, but their nature can easily be demonstrated by the addition to the specimen under the microscope of a drop of iodine in iodide of potassium (p. 33), the development of a deep purple colour, turning to black when a drop of sulphuric acid is also added, proves the presence of starch.

If only a milk diet has been given, as in infants, the

appearance is different.   The stools are loose instead
of formed; the constituents described above are absent,
with the exception of fat globules and crystals, these
being present in large numbers.   Coagulated albumen
(casein) is also found, usually as irregular yellow
particles.

Crystals.—In the microscopic examination of fæces
several varieties of crystals will be met with but very
few of them have any diagnostic value.

Fat crystals (see above).—They occur very abun-
dantly in the stools of patients suffering with jaundice.
The investigations of Gerhardt, v. Jaksch and Oester-
lein, show that they are formed of a combination of the alkaline earths with the higher fatty acids.   The crystals
take the form of minute, sharp-pointed crystals, scattered through the field or arranged in clusters like
rosettes.

Hæmatoidin crystals.—The author has never
found these crystals in the alvine discharges, but von
Jaksch, states (*l. c.*, p. 156) that he has not infrequently
done so, especially in chronic intestinal catarrh from
over-eating, and in many instances in which blood had
been discharged into the intestine some time before the
stools were passed.   The form of the crystals is usually illdefined; they are sometimes free and sometimes
enclosed in masses of a shining substance resembling
mucin.

Charcot-Leyden crystals (for an illustration
of these bodies see fig. 54).—As these crystals are
the result of decomposition of albuminous substances,
it is only to be expected that they are sometimes found in the stools.   They have, however, no diagnostic significance and have been found in various
diseases.

Oxalate of lime crystals (fig. 37, p. 199) are frequently met with and are probably derived from the
food, being particularly abundant after a vegetable diet.
Other crystals occasionally found, but having no clinical
value are :—triple phosphate, phosphate of lime, carbonate of lime and sulphate of càlcium.

The **formed elements** which are found in the
motions are blood corpuscles, white and red, and epithelial cells, but the most important from a practical
point of view are the various parasites with their ova,
and some pathogenic micro-organisms. The same
plan will be adopted in their description as in the
chapters on the examination of sputum, urine, vomit,
&c.

Red blood corpuscles.—The occurrence of blood
in the stools has already been briefly alluded to (p. 216),
but although the colouring matter (somewhat altered by
the action of sulphuretted hydrogen) is frequently present, yet the red blood cells are very seldom seen, even
when the blood is apparently abundant, as in typhoid
fever. Small irregular particles of hæmatoidin, reddish-
brown in colour are met with, and more rarely, crystals of this substance. The presence of blood may be
demonstrated thus :—A small portion of the excrement
is dried, and the powder placed on a glass slide ; a
crystal of common salt is then added, together with a
few drops of glacial acetic acid. The specimen is
covered with a cover-glass, and heated until steam just
begins to rise; after cooling, minute crystals of hæmatin
will be visible by aid of the microscope.

Leucocytes.—White blood cells are but seldom
met with, even when there is considerable inflammation
of the intestine. If ulceration takes place, pus cells
may be found in considerable quantity. The discharge

Q

of an abscess into the bowel is indicated by the appearance of pure pus.

Epithelium.—Epithelial cells in greater or less
abundance are always found in the fæces, and their
presence therefore is of very little clinical importance.

Two chief varieties are met with, squamous cells
from the lower end of the rectum, and cells which are
fusiform or columnar in shape, from the remainder of
the intestinal canal.   These latter are sometimes difficult to define, their outlines being faint, but their nuclei
are always seen, and are especially distinct if some
dilute staining fluid be added, such as methylene blue
or magenta.

Nothnagel has drawn especial attention to the fusiform cells, which he considers a degenerative form,
caused by the abstraction of fluid.

In addition to the above constituents of the evacuations, a certain amount of amorphous material is always
present, derived chiefly from the waste products of digestion ; it is of little importance medically, either with
a microscopic or on chemical examination.

The subject of **parasites** discharged from the
bowel is a very large one.   It will be most convenient
to follow the usual division into vegetable and animal
parasites, subdividing them into those which are pathogenic and those which are not.   The former being the
most important will be considered first in each division.

Vegetable Parasites.

Pathogenic :—

Bacillus of tubercle.—Tubercle bacilli are frequently found in the stools of phthisical patients af-

flicted with diarrhœa, and in such cases ulceration of the intestines is almost invariably found after
death. It is of course conceivable that the bacilli
may be derived from the sputum which has been swallowed, but if such were the case, they would be very
few in number and it would be no proof that an intestinal lesion did not also exist, as ulcers are frequently
found after death, in which diarrhœa was not a prominent symptom during life.

It is not, however, with obviously consumptive patients that an examination of the fæces for tubercle
bacilli is of value, but in those cases in which there is
obstinate diarrhœa, with raised temperature, and in
which the diagnosis is doubtful. If the organisms are
then found the uncertainty is cleared up, and if they are
very plentiful so as to resemble almost pure cultures,
ulceration is nearly certain to be present. In order to
search for tubercle bacilli, a small portion of the fæces is
placed between cover-glasses, which are dried, passed
three times through the flame and stained in the same
way as will be described in the chapter dealing with the
examination of sputum.

Bacillus of typhoid fever.—The pathogenic
value of the bacillus first described by Eberth in 1880 in
causal relationship with enteric fever is now accepted
by most bacteriologists. The bacilli, however, unlike
those of tubercle, possess no characteristic staining property; it is therefore impossible to demonstrate them
satisfactorily in typhoid stools, although they may
readily be demonstrated in the tissues after death by
the process described on p. 91 (Löffler's method).

To prove their presence in the fæces a troublesome
and prolonged process of separation of the colonies by
plate cultivations has to be adopted, and this can only

be accomplished in a properly appointed bacteriological laboratory.

Bacillus of cholera.—The discovery by Koch of the "comma bacillus," which has now been conclusively proved to be the cause of Asiatic cholera, was the chief means of placing bacteriology in the position which it now holds in modern science. Fortunately in England at the present day, we have very rare opportunities of studying the cholera bacillus except in laboratories. Koch found the organism in the intestine and alvine discharges of patients who had died or were suffering from Asiatic cholera. It was rarely found in the vomit, and never in the other secretions, urine, saliva, &c., nor in the blood or breath.

The bacillus is a short thick rod, not so long as the tubercle bacillus, and generally more or less curved. The organisms are often united together into wavy threads (spirilla), especially in artificial cultures, owing to the rapidity of reproduction. As with the bacillus of enteric fever, the cholera organism is very difficult to demonstrate in the stools owing to the myriads of other bacilli which are present, but as in the typical rice-water stools it often occurs in huge numbers, an attempt may be made to do so in the following manner:—

A small portion of the flaky sediment of the stools is placed on a cover-glass, which is then covered by a second, so as to obtain two uniform layers by gently sliding the glasses apart. They are dried and the films fixed by passing them three times through a flame. They are then placed, prepared side downwards, in a watch-glass containing an alcoholic solution of fuchsine or methylene blue. After five minutes they are removed and thoroughly washed in water, after which they are dried and mounted in balsam.

More satisfactory results are obtained by adopting Koch's method of plate cultivation, but this requires considerable experience in bacteriology. It may, however, be briefly stated here that cultivated on plates of nutrient gelatine, the colonies commence to form in about twenty-four hours; they are white in colour, roughly circular in shape, but have irregular outlines. Gradually a yellowish tint appears, which grows darker, especially towards the centre and the gelatine begins to liquefy, emitting a most unpleasant odour.

For further descriptions of this bacillus and its growth in other media, the reader is referred to the text-books on bacteriology.

Bacillus of cholera nostras.—This organism is often known as the "bacillus of Finkler and Prior," these observers having first described it. It is found in the stools of patients afflicted with "cholera nostras," and may be demonstrated in the same way as the bacillus of Asiatic cholera; it differs from the last named in being broader and larger, but the most marked differences are seen in the cultures on nutrient gelatine. The colonies are very much larger than those of the comma bacillus of the same age. They have a faint yellowish-brown tinge, the borders are well-defined, and they exhibit a distinctly granular appearance. They liquefy nutrient gelatine very rapidly, so that the first plate of a series may be completely liquefied on the day following inoculation, emitting a very foul odour.

NON-PATHOGENIC ORGANISMS.

Micro-organisms of various kinds abound in the fæces, both in health and disease. So numerous are they that they often form a considerable proportion of the whole bulk of the stools. It is on account of this that pathogenic organisms such as those of typhoid or dysentery are almost impossible to recognise, not possessing any characteristic selective powers for particular stains.

All the fission fungi stain readily with the ordinary stains, fuchsine and methylene blue being the best.   A small portion of the discharge is spread out into a uniform layer between cover-glasses, which are then separated by a sliding movement, as in the examination of sputum, dried and passed three times through the flame.   The glasses are then placed face downwards in a watch-glass containing the undiluted stain, or a drop or two of the dye is placed on the glasses.   After three minutes, the preparations are washed thoroughly in water ; if fuchsine be used, a few drops of methylated spirits may be added.

Very excellent specimens may be obtained by means of the Gram method.   After the glasses are prepared in the ordinary way, a few drops of methyl-aniline-violet are placed on them and allowed to remain for three minutes, they are then washed in Gram's solution (p. 33) for one minute, and are finally dipped alternately in aniline oil and methylated spirits until no more colour comes out, after which they are dried and mounted in balsam.   Another method by which good results are obtained, and which, as will be presently again mentioned, is strongly recommended by v. Jaksch, is to immerse the glasses for about a minute into a solution of iodine in iodide of potassium.   This reagent has the advantage of staining the yeast fungi as well as bacilli and micrococci, and for some of the latter it has a peculiar power of differentiation, colouring them blue or violet.

Moulds.—Members of this group are very seldom found in the fæces when first passed. Practically the only one ever met with is the thrush fungus—Oïdium albicans, which consists of small round spores and mycelial threads. It is generally found in the stools of children suffering from a similar condition in the mouth. It is of very slight diagnostic importance.

Saccharomycetes.—These are very commonly found in the alvine discharges, both in health and disease, occurring as round or oval cells, exhibiting the characteristic gemmation ; they are especially abundant in the acid stools of children. They are also found in large numbers in acute catarrh of the small intestine in adults. When a cover-glass preparation is stained with iodine solution, the cells assume a mahogany-brown colour.

Schyzomycetes.—The fission fungi often occur in such large numbers as to constitute the main bulk of the discharge. In sprue, for instance, the peculiar whitish scum found in large quantities in the stools in the exacerbations of the disease, consists entirely of bacilli of various sizes. Dr. Thin has isolated thirteen varieties, probably one of these is pathogenic, but neither Dr. Thin nor the author has been able to verify this assumption.

The organism most frequently met with is the one associated with ordinary putrefaction, namely, the Bacterium termo.

Nothnagel was the first to describe the Bacillus sub-lilis in the stools. It has no pathological value. It occurs as long mycelial threads, the spores being situated at the end of the rods, or attached to the threads. These micro-organisms stain brownish-yellow with the iodine solution.

Von Jaksch in a most exhaustive investigation on this subject, has described a large number of fungi which stain blue with the iodine solution, none of which, however, have any clinical value (" Clinical Diagnosis," p. 132). Some of them are micrococci, which occur in large numbers, and are coloured reddish-violet with the solution.

Next in order are small bacilli resembling those of mouse septicæmia, these stain in a similar manner.

Larger rods are sometimes seen, which resemble the Leptothrix buccalis. Finally, large round cells are met with, strung together like beads, appearing very much like yeast cells.

It may be here remarked that the colouring produced by the iodine solution is only transitory and has usually completely disappeared in forty-eight hours.

Animal Parasites.

Protozoa.—Various forms of monads and infusoria are met with in the fæces, but are comparatively uncommon, and merit only passing notice.

Nothnagel found monads in the stools of consumptive and typhoid patients ; active movements had generally ceased.

The **Amœba coli** was first described by Lösch (" Virchow's Archiv.," 1875, p. 196). These bodies are circular in form, contractile, and consist of coarsely granular protoplasm, and possess a round nucleus.

Some valuable researches on amœbic dysentery have recently been made by Drs. Councilman and Lafleur (" Johns Hopkins Hospital Reports," December, 1891, vol. ii., p. 395, *et seq.*). They found in certain cases of dysentery amœbæ in the stools which did not differ materially from those described by Lösch. In searching for the organisms some choice had to be exercised in the portions of stools to be examined. They were more numerous in the small gelatinous masses which were often contained in the fæces than elsewhere. The numbers which were found in different cases and even in the same case at different times, varied greatly. At times no single portion of the fæces which could be examined was free from them. In other cases single individuals were only found after long and careful search. Their numbers in general were proportional to the severity of the lesions.

Drs. Councilman and Lafleur state that the amœbæ differ somewhat in appearance according as they are active or inactive. In the latter condition they are round or slightly oblong and are more highly refractive than the other cells which may be found in the stools. In this resting condition there can frequently be distinguished no division into an ecto- and an endo-sarc, but there is simply a body, enclosing vacuoles of greater or less size. They are difficult to recognise in this condition unless active ones are found at the same time and so attention be directed to their presence, but when in motion their appearance is characteristic. Every degree of activity of movement can be seen. At times the change is so slow that one must look carefully for some time to be sure that any motion is taking place and at others the movements are so rapid that certain of their phases can scarcely be followed. The movements consist essentially of two kinds: one progressive, and the other limited to the thrusting out and retraction of pseudopodia. In some instances, there seems to be a certain rhythm in these changes of form and position, which occur at regular intervals.

Various bodies were frequently enclosed in the amœbæ. The most common of these were red blood corpuscles, often in considerable numbers. Pus cells, well preserved, or in various stages of disorganisation were also seen. Both bacilli and micrococci were occasionally enclosed. In some cases, in addition to red blood cells, blood pigment was found, either in the form of black granules or as irregular brown masses.

Of **infusoria** the most common is the **Cercomonas intestinalis.** It occurs in the mucous discharges of young children. It is pear-like in shape, furnished with tentacles, and has a clear nucleus.

An organism, closely resembling the one first described, is the **Trichomonas intestinalis,** but distinguished from the former by a ciliated disk at one end.

Another entozoon occasionally found, especially in diarrhœa, is the **Paramæcium coli,** it is oval in shape and covered with cilia, and the anterior extremity is narrower than the posterior.

VERMES.

The recognition of the various worms which infect the human intestinal canal is of great importance, not only with a view to undertake treatment, which may expel them, but also as a means of clearing up vague symptoms, which have previously evaded explanation.

It is not intended here to enter into the full life history of each parasite, and to trace it from the ovum through the "intermediate host" to the human intestine; but only those forms will be described which are likely to come under the notice of the medical man in his clinical work.

In addition to the worm, the ova are sometimes found in the stools ; these are never discharged through the genital canals of the parasite, but remain *in situ* until the proglottis is ruptured.

CESTODA.

The tape-worm as a whole is known as a " strobilus," and is divided into a " head," " neck," and a number of segments or proglottides.

Tænia mediocanellata (saginata).—This is the most common tape-worm in this country. The intermediate host is the ox.

The head, which is the most easily recognised part of the parasite (fig. 44), is square in shape, and measures in the transverse diameter 1·5 mm. It is provided with four suckers, which usually have much pigment deposited round them.

As the head is generally difficult to find, or may not come away when the rest of the strobilus is discharged, it is important to be able to recognise the proglottides. The whole worm may attain the length of four yards, and acquires its full growth in about three months, so that if the head has not been discovered, although proglottides have been passed, we may conclude that the parasite has been destroyed, if no fresh segments are discharged for thirteen weeks.   There is no distinct neck, but the first segment appears immediately below the head.   The upper segments are broad and short ;

about the centre of the animal they become square, and in the lower half the length exceeds the breadth.

The sexual organs attain their full development about the 450th joint from the head, the ova appearing about 400 joints lower down.

The ripe proglottides measure $\frac{3}{4}$ inch in length, and about $\frac{1}{3}$ inch in breadth.   They usually rupture and discharge their ova while in the intestine, so that those

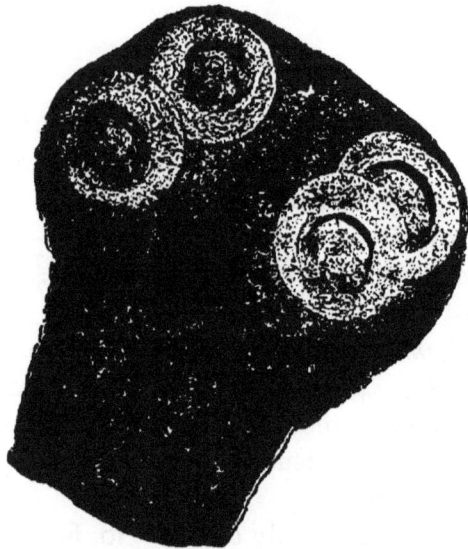

FIG. 44.—Head of Tænia mediocanellata (Boyce).

which are passed per anum are shrivelled and empty (Fagge).

The segments may be recognised and distinguished from those of the other tape worms by the form of the uterus. This organ has numerous branches, having twenty to thirty on each side of a median channel; the branching is dichotomous, and the terminations are round and club-shaped. The genital pore does not alternate regularly, but may be situated on the same side in two or three successive segments, they are, however, always situated on the narrow edge.

Malformations are particularly liable to occur, and sometimes there are two or three genital pores in a single proglottis, each corresponding with a separate double sexual apparatus; sometimes the segments are not regularly developed, so that imperfect triangular joints disturb the symmetry of the series; or a supernumerary proglottis projects by the side of the continuous line of joints; very rarely there are two distinct chains united in their whole length by one edge at an acute angle.

The eggs are oval, and exhibit the primordial yolk membrane; no hooklets are seen in the embryo as is the case in Tænia solium.

Tænia solium.—This worm was so named because, according to an old idea, it was thought to exist alone in the intestine. This is now known not to be the case; sometimes they occur in large numbers, as many as five-and-twenty having been passed by one individual. It was also formerly supposed to be the most common of the tapeworms in this country, but this, too, has been shown to be without foundation, the Tænia mediocanellata occurring far more frequently.

The head of the Tænia solium (fig. 45) is of a generally rounded form, about the size of a pin's head. In front it is prolonged so as to form a proboscis, or rostellum, which is surrounded by a circle of twenty-six hooklets arranged with their points outwards. They are of two sizes, large and small alternately. The wide part of the head is provided with four large sucking discs, and is generally black from the presence of pigment. Owing to its small size it is difficult to find, and is frequently overlooked by the patient, who should be directed by the physician to bring for examination every particle of white matter passed.

The loss of the head is rendered more probable by the very slender neck which it surmounts. This neck is about $\frac{1}{2}$ inch in length, and it gradually merges in the anterior part of the body, in which fine transverse lines begin to appear as the first indication of the formation

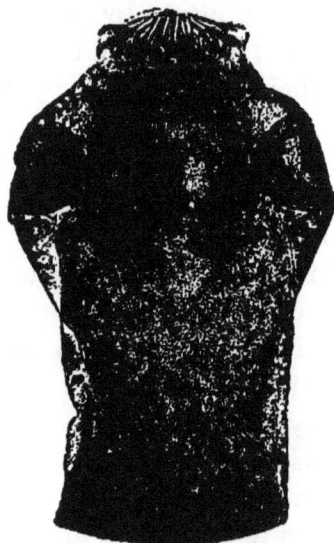

FIG. 45.—Head of Tænia solium (Boyce).

of segments. On passing down, the segments elongate and become more completely divided. The whole length of the strobilus is usually about nine feet. The joints are at first very small, and broader than they are long. They gradually increase in breadth and length until about the middle they appear square, and in the lower half they are longer than they are broad.

The proglottides may be distinguished from those of

the mediocanellata by the shape of the uterus. This is more simple in the solium, consisting of a central passage running parallel to the length of the proglottis, and having about ten branches on each side, which ramify instead of dividing dichotomously. The genital pore is placed in a little papilla, situated on the lateral edge of the segment. It alternates regularly on the two sides of the animal throughout its whole length.

Proglottides are best demonstrated by placing two of them on a glass slide and covering them with a large cover-glass, when they may be examined with a good lens. The specimens will be rendered more transparent by the addition of a little glycerine.

The total number of segments is about 800. The sexual apparatus begins to appear about the 200th segment from the front, and is mature about the 450th. Its time of growth is the same as for the mediocanellata, and the same deductions may be drawn from the appearance of the segments.

The ova (fig. 46, g) are globular, and measure ·036 mm. in their longest diameter. They have a shell which appears to be marked with a number of minute radiating lines, this being due to rod-shaped projections, which closely cover its surface; when mature, embryos furnished with hooklets are seen within the eggs. The intermediate host of this parasite is the hog, hence its greater frequency on the continent, where uncooked pork is a frequent article of diet.

Tænia nana is the name given to a form of tapeworm prevalent in Sicily, Italy, and Egypt. Ranson mentions it as having occasionally been found in England. Its average length is nearly $\frac{1}{2}$ inch, and its breadth about $\frac{1}{50}$ inch. It has a circular head provided with four suckers, and a rostellum surrounded by about twenty hooklets. The uterus is oblong. This parasite occurs in huge numbers in a single patient.

Bothriocephalus latus.—This worm is limited to the inhabitants of certain countries of Europe, being most commonly found in Switzerland and Sweden. It is the largest of all the tapeworms, attaining a length of from 16 to 20 feet, and possessing from 3000 to 4000 segments, which are broader than they are long. In the middle of the strobilus they are about $\frac{1}{2}$ an inch broad by $\frac{1}{4}$ inch in length, but the last ones gradually become longer and narrower so as to be almost square. This worm has a longitudinal projecting ridge traversing its whole length. The uterus occupies the middle of each segment. It is composed of a convoluted tube bent several times upon itself so as to give a rosette-like appearance. The genital pore lies in the centre of each proglottis, opening upon its ventral surface. When portions of the worm are passed by a patient, the segments do not come away separately, but portions two and three feet in length are expelled. The head of this parasite is oval, and about $\frac{1}{25}$ inch in breadth. It is blunt at the anterior extremity, and has two deeply grooved longitudinal suckers, one on each side.

The eggs (fig. 46, *d*) are .oval, and are about ·07 mm. in length. They are covered with a brown shell, and open in a peculiar manner at one extremity by a kind of small lid, so allowing the embryo to escape.

The intermediate host of this tapeworm is not known, but is supposed to be some fish. The frequency of fresh water lakes in Switzerland may explain its prevalence in that country.

Other tapeworms which are occasionally found in the human subject, but are so rare as only to need mention here, are the *Tænia cucumerina* which is about 8 inches long, having a small head with a rostellum surrounded with a quadruple circle of hooklets; the *Tænia flavopuncta;* the *Tænia madagascariensis,* and the *Tænia leptocephala.*

The ova of the above Cestodes are occasionally found in the fæces without any segments accompanying them. Should the presence of a tapeworm be suspected, the stools should be mixed with water and the sediment allowed to subside, and this process should be repeated several times until very little matter remains; a small portion of this should be placed on a glass slide, a cover-glass applied, and the specimen examined by a ¼ inch lens, when the eggs, if present, will be easily recognised.

The same process should be adopted in searching for the smaller parasites now to be described.

TREMATODA.

These worms are commonly known as "Flukes," they are flat and more or less oval in form. On their ventral surfaces they have one or more sucking discs. They are rare in man, but have occasionally been found in the stools, or more often in the biliary passages or intestines.

Distoma hepaticum.—This worm is very common amongst sheep, producing the disease called "Rot." As its name implies its chief habitat is the liver. It frequently dilates and obstructs the bile ducts. It is about an inch in length and half an inch in width. The body is flat and is elongated anteriorly, forming a kind of head which is furnished with a sucker. There is another sucker on the ventral aspect. Between the two lies the genital pore leading to the uterus which is

R

convoluted.   The eggs (fig. 46, *e*) are small oval bodies,

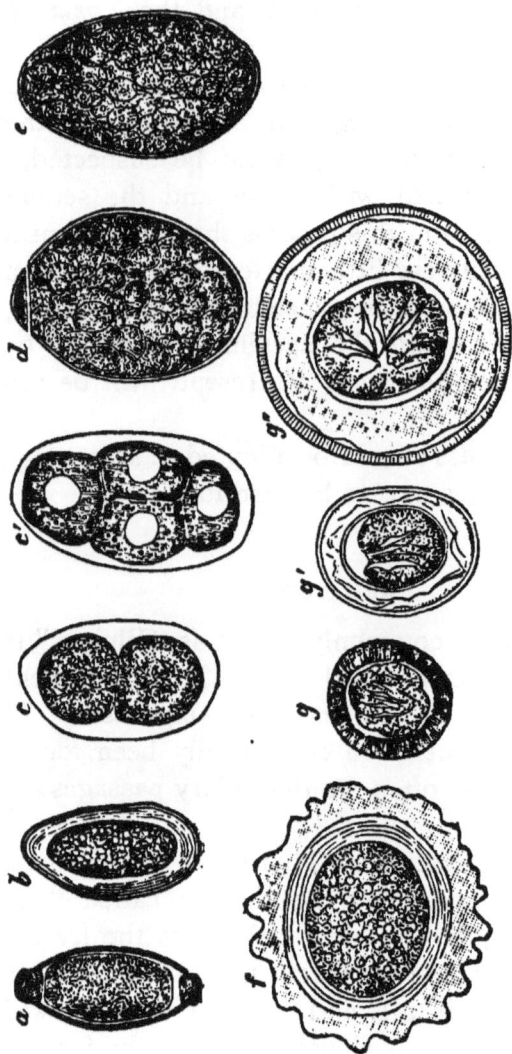

Fig. 46.—Ova of intestinal parasites (Bizzozero and Firket).   *a.* Trichocephalus dispar.   *b.* Oxyuris vermicularis.   *c.* Anchylostoma duodenalis.   *c'.* The same, in process of segmentation.   *d.* Bothriocephalus latus.   *e.* Distoma hepaticum.   *f.* Ascaris lumbricoides.   *g.* Tænia solium.   *g' g''.* Eggs, the source of which could not be ascertained, probably large eggs from one of the Tæniæ.

·13 mm. long; they are brown in colour, and one end which is broader than the other opens by a small lid.

Distoma lanceolatum.—This species is about a ¼ inch long and 1/15 inch in breadth. It is pointed at the extremities, but more so anteriorly than posteriorly. It has no alimentary canal, but possesses a complicated generative apparatus packed with numerous ova. These are oval, brown in colour, and about ·04 mm. in length. This trematode is very rarely seen in man, and only occurs in comparatively small numbers in cattle.

Distoma sinense.—This parasite is more elongated than the two just described. It is about ½ inch in length and about ⅛ inch in breadth. It possesses a single suctorial disc at one extremity, communicating with a bifurcated intestine; the genital pores open at the opposite end. When seen in the bile in the fresh state, a beautifully delicate green colour, tinged with yellow, may be observed round the edges, whilst the centre of the animal is a deep brown. The ova are extremely small and occur in very large numbers. This worm only occurs in man, and as many as five hundred may be found in the bile ducts of one person. Hitherto it has only been described amongst the Chinese.

NEMATODA (Round Worms).

Ascaris lumbricoides (common round worm).— The Ascaris lumbricoides inhabits the small intestine, but may be passed either alone, or with the stools. In general appearance it closely resembles an ordinary earthworm. Whilst living it has a reddish-brown colour, fading after death to a duller grey. It measures six to sixteen inches in length, is marked by transverse striæ, and tapers towards both extremities. The female is about twice as long as the male the head; consists of

three conical projections which are furnished with fine
teeth and touch corpuscles.   The posterior extremity of
the male is folded on the anterior surface like a hook.
The ova (fig. 46, f)are circular or bluntly oval; yellowish-
brown in colour, and measure ·07 mm. in length.   They
are covered with an albuminous sheath which is often
tinged with bile.   Beneath this covering is a dense
shell.   Several of these parasites (three to four hundred)
may be passed by one individual ; but probably each
worm only exists for a few months within the body of
its host.

Ascaris mistax.—This parasite in general form
much resembles the preceding, but is smaller, and the
shape of its head is rather more pointed.   Hitherto it
has only been found in the cat.

Oxyuris vermicularis (threadworm).—As their
common name indicates these worms are like small
ends of white thread ; the male being about an eighth
of an inch and the female about four-tenths of an inch
in length.   They have a comparatively blunt anterior
and a tapering caudal extremity.   The tail of the male
is provided with six pairs of papillæ.   The male is much
more frequently seen than the female.

The eggs (fig. 46, b) are oval in form and flattened on
one side.   They measure ·05 mm. in length.   The contents are granular except when mature, when the embryo may be seen.   The shell is membranous and presents a treble contour.

Trichocephalus dispar (whipworm), (fig. 47).—
This parasite inhabits the cæcum.   It is cylindrical in
form, having a short thick body, and long whip-like
neck.   The male is about 1½ inches in length, and the
female about 2 inches.   The worm fixes its thin extremity by means of four claw-like teeth into the wall

of the intestine, and is therefore only very rarely found in the fæces. The ova (fig. 46, a) are oval and about ·05 mm. long, and are characterized by having a translucent point at either extremity.

Anchylostoma duodenale (Strongylus duodenalis).—This parasite does not occur in this country, but is frequently met with in Egypt and Brazil. It inhabits chiefly the jejunum. It is cylindrical in form. The male measuring about half an inch, and the female an inch in length. The head which is bent nearly at right angles to the body is armed with four conical teeth, by means of which the animal fixes itself on the mucous membrane. The posterior extremity is tapering, and in the male expands into a pouch with three flaps. The eggs (fig. 46, c) are oval, with a thin transparent shell,

FIG. 47.—Trichocephalus dispar.

and have a long diameter of ·05 mm. The division of the yolk (fig. 46, c') can be plainly seen. This nematode is only seen in the stools after some measures have been taken to expel it, but the eggs frequently appear.

Strongylus gigas.—This is a large worm chiefly met with in the kidney, bladder, lungs and liver of dogs, but a few cases are on record in which it has been met with in the pelvis of the kidney in man. It is a large worm, the male being about ten inches in length, and the female nearly a yard.

Anguillula intestinalis.—This is a small round worm, which has been discovered in the stools in cases of Cochin-China diarrhœa, but has not been met with in this country.

Trichina spiralis.—These parasites have in very rare cases been found in the stools. They closely resemble the Trichina found in muscle, but are not encapsuled. The male is 1·5 mm. in length, and the female 3 mm. The males are distinguished from the females by having two conical projections from the posterior extremity.

CHAPTER XVI.

EXAMINATION OF SPUTUM.

By "sputum" is understood all the matter coughed or hawked up from the air passages. As this material includes the secretions from the buccal respiratory tract, it might naturally be expected that an examination of the contents of the spittoon would yield valuable results as regards the condition of the parts forming that tract. Since the discovery of the tubercle bacillus by Koch, in 1882, the microscopic examination of sputum has received a great impetus, and it has now become almost a matter of routine in any case of lung disease of a doubtful character.

The naked eye features of the expectoration received much more attention from the older writers, before the physical examination of the chest by percussion and auscultation arrived at the present state of exactness, and before the clinical thermometer had been brought into routine use. Not only were the appearances of the secretion noticed, but attention was directed to its smell, taste, reaction, &c. With our modern methods of physical examination, much less stress is now laid on the macroscopic characters of sputum, but with the aid of the microscope a diagnosis, which may be doubtful, or extremely obscure (*e.g.*, Actinomycosis), may be corroborated or established.

Although this chapter will necessarily be directed chiefly to the microscopic examination of sputum, it will

be necessary, or at any rate desirable, to briefly consider some of the more noticeable points in connection
with its quantity, colour, odour, &c.

Sputum may be described as mucous, muco-purulent,
or purulent, according as it is transparent, semi-opaque,
or resembles the matter drawn from an abscess. To
attempt to describe its examination under these heads
would be manifestly absurd, as instances of each are
found in the course of most diseases of the respiratory
organs.

It is very important that a satisfactory sample should
be obtained. In the first place then, the spittoon must
be carefully washed out before use, and must never be
made a receptacle for grape skins and various other
odds and ends, which seem so frequently to find their
way into it. It is well also to replace it with another
during meal times, otherwise particles of food are certain to become mixed with the expectoration, which
gives much unnecessary trouble to the physician when
searching for the minute white particles so characteristic
of certain diseases. This is especially the case if the
patient be on a milk or farinaceous diet. Thus, in the
sputum of one case which I had sent me for examination in several succeeding samples, minute white specks
were seen, which combined with the history and symptoms of the case had given rise to a suspicion that
it was one of actinomycosis. By means of micro-
chemical tests, however, they were shown to consist
simply of collections of starch grains derived from food.

Sputum coughed up early in the morning before the
patient has partaken of any food is to be preferred,
though for some purposes the whole quantity collected
during twenty-four hours is necessary.

In order to observe the stratification peculiar to some

diseases, the sputum should be collected in conical glasses.

QUANTITY.

The amount of sputum expectorated within twenty-four hours varies very greatly. In some conditions the spittoon may be filled several times daily, whilst in others there will scarcely be a drachm. The sputum is excessive in œdema of the lungs, and in some cases of hæmoptysis; also when an abscess bursts, derived either from the lung itself or breaking through into the lungs from neighbouring parts. Most commonly this is due to the rupture of an empyema, but the pus may be derived from the abdominal cavity. The quantity is also excessive in bronchiectasis, the patient may rest for several hours, suddenly bringing up many ounces of fœtid purulent matter.

In phthisis again, where large cavities exist, the expectoration is often very copious but this is by no means characteristic.

CONSISTENCE.

The consistence of the sputum stands usually in some relation to its quantity, for if this be very copious the expectoration is generally thin and watery, whilst if small in amount it is usually thicker; but numerous exceptions to this statement are constantly met with. The consistence is dependent in a great measure upon the amount of mucoid material contained in it. As a type of a tough tenacious sputum, that occurring in croupous pneumonia may be taken; it is often so tenacious that if the spittoon be turned upside down, it will not flow out. The reason for this is difficult to assign, for the proportion of water is not small, nor is the quantity of mucin contained nor its specific gravity unusually large. The expectoration in bronchial asthma is often very tough, as can be proved by the difficulty experienced in obtaining a thin layer for microscopic examination ; it contains a large amount of mucin. In the early stages of acute bronchitis also the secretion is very thick and tenacious. This condition was termed by the older writers Sputum crudum, in contra-distinction to the thin watery expectoration which they named Sputum coctum.

COLOUR.

In speaking of the colour of sputum we must consider briefly its *transparency*. Pure mucoid expectoration is colourless and almost transparent. The more cellular elements there are present, especially leucocytes, the more cloudy and opaque the secretion becomes.

Changes in colour are almost entirely due to the presence of blood or its derivatives.

This is not the place to enter into a discussion on the different causes of hæmoptysis. Roughly speaking, they may be divided into two classes, in the first of which a rupture of a blood vessel, either in the pulmonary or systemic circulation, gives rise to the hæmorrhage, whilst in the other there is a diapedesis of red blood corpuscles—parenchymatous hæmorrhage. The amount of blood may vary from fine streaks distributed through the expectoration to almost pure blood. The latter condition is only found when a vessel has completely ruptured. Investigations by Dr. Kidd and Mr. Taylor at the Brompton Hospital for Consumption showed that in a large proportion of the cases of profuse hæmoptysis, a small aneurysm had burst, whilst erosion of the vessels was comparatively rare. When first ejected the blood is usually bright in colour, frothy in appearance and alkaline in reaction.

Hæmoptysis is by no means diagnostic of tubercle; heart disease, especially mitral stenosis, bronchitis and pneumonia, croupous and catarrhal, congestion of the lungs, infarcts and the presence of parasites, new growths, &c., all may give rise to blood spitting, more or less profuse. Professor Albert Fränkel in a most exhaustive discussion ("Pathologie und Therapie der Krankheiten des Respirations Apparates," vol. i., p. 110) on the occurrence of hæmoptysis, comes to the conclusion that blood spitting without any previous symptoms, is most probably always the expression of commencing tubercular inflammation.

If the blood has been retained in the lungs for any time it undergoes changes which produce considerable alteration in its colour and general appearance, this being due to decomposition, or metamorphosis of the hæmoglobin. Thus in gangrene of the lungs, the expectoration becomes of a dirty **reddish-brown** colour, owing to the formation of methæmoglobin and hæmatin.

Very characteristic also is the peculiar **chocolate** colour which the expectoration assumes when a croupous pneumonia is passing into gangrene. This form was termed by Andral "prune juice expectoration." Under the microscope the blood corpuscles will be found to have undergone very considerable changes, which will again be referred to later on. The same appearance is sometimes noticed in the sputum of aged persons suffering from œdema of the lungs. A peculiar **brownish** colour of the expectoration is often found in gangrene of the lung more commonly when ulceration of the lungs has occurred from some other cause. Leyden considered this to be due to the abundance of hæmatoidin and bilirubin crystals which are present.

Brown induration of the lungs produces a peculiar appearance in the sputum. In the mass of mucoid material **rust coloured** streaky deposits are seen, especially in the early morning, and these under the microscope are found to consist of pigmented epithelial cells.

More common than the above is the well known rust-coloured sputum of croupous pneumonia. This colour, however, is not always constant, occasionally it assumes a **lemon-coloured** hue, or more commonly a **grass-green** colour, this being especially the case when the crisis has been deferred. This latter appearance must not be confused with those serious cases in which jaundice has supervened, for the green colour is then due to admixture with bile colouring matters.

Traube also described this green colour in sub-acute cases of pneumonia, so that in any doubtful case when green expectoration is observed, micro-organisms should always be sought for.

Various observers have described a **yellow ochre** coloured sputum ; this being associated with abscess of the liver, more especially when hydatid cysts are present. Under these circumstances large numbers of hæmatoidin and bilirubin crystals are seen under the microscope and the hooklets of the parasite may be discovered on microscopic examination.

The above colourations must not be confounded with

the **yellow tint** sometimes assumed by the expectoration in summer time. This occurs in various diseases and the tint is very similar to the yellow of an egg. The colour is confined to the superficial layers of the mucus, and often becomes more pronounced after the expectoration has remained some time in the spittoon. It is dependent upon the presence of bacteria, which in their growth produce the pigment. It is a fact worthy of remark that not infrequently several patients in the same ward exhibit the same colouration of the sputa, especially those in neighbouring beds. From similar causes a greenish tinge is sometimes observed.

Black coloured sputa may be said to be normal, for they are almost always met with in towns or manufacturing districts. The black hue is owing to the admixture of particles of carbon, dust, &c., which everybody inhales more or less. The particles are found partly free and partly embedded in cellular elements, hereafter to be described.

In cases of malignant disease of the lungs the expectoration is generally tinged a **dark red,** and has a peculiar gelatinous consistency, sometimes, however, both with carcinoma and sarcoma, various shades of **green** appear.

Before quitting the consideration of the diagnostic value of the colour of the sputa, a few words must be said about what is known as "spurious hæmoptysis." That is to say blood ejected from the mouth, but having its origin not from the lungs, but from the upper part of the air passages, the trachea, pharynx, mouth and nose.

The diagnosis from lung disease, however, is more easily made clinically than by an examination of the contents of the spittoon.

Hysterical patients often prick and suck their gums

in order to elicit sympathy and obtain credit for being the subject of serious pulmonary disease.

Dr. Douglas Powell ("Diseases of the Lungs and Pleuræ," 3rd edit., p. 363, *et seq.*) divides false or spurious hæmoptysis into seven divisions:—1. In cases of epistaxis, the blood commonly trickles down the back of the throat and excites cough, by which it is removed in clots, staining the saliva. 2. Ulceration of the throat, especially when malignant, may lead to copious hæmorrhage. 3. Hysterical hæmoptysis (already referred to). 4. A morbid state of the gums frequently arises from want of due attention to the teeth, or from the presence of decayed stumps in the alveoli; blood may exude on the slightest friction. 5. An insufficient supply of vegetable food leads to a spongy congested state of the mucous membrane of the mouth and fauces, very similar to the lesions characteristic of scurvy, and this is one of the most common causes of spurious hæmoptysis. 6. In certain cases of anæmia the mucous membrane of the mouth and fauces exudes a sanguineous fluid. The transudation is very slow, and in the daytime scarcely noticed, but during the night some accumulation takes place, and on waking the patient expels some bright red, unaërated fluid containing a few coagulated films, giving an appearance closely resembling that of currant jelly and water. 7. General hæmorrhage from the whole mucous membrane of the mouth is sometimes seen in hæmophilia.

Odour.

Inferences to be derived from the odour of the sputa are not of any material value. The chief exception is that of gangrene of the lungs, in which complaint the expectoration has a peculiarly offensive and penetrating smell.

In phthisis and other diseases in which the secretion is retained for some time in the air passages, it acquires a faint sweetish odour, due to decomposition.

A few other naked-eye characters of the sputum remain to be considered.

In the later stages of phthisis, when cavities have formed, the sputum becomes **nummulated.** It then consists of opaque greenish discs, about the size of a sixpenny piece. If ejected into water these discs sink and spread out into irregular and ragged masses. They are probably formed by the passage of a thick secretion, through the very fine bronchial tubes (Fränkel). A similar condition is seen when an empyema has burst into the lungs (Traube).

When some forms of sputa are allowed to stand they separate into three **layers,** the uppermost of which consists of semi-opaque, mucoid, and generally frothy liquid. The middle is a serous-like fluid, somewhat cloudy and whey-like; whilst the undermost is made up of a mass of pus cells and débris, and is usually of a dirty yellowish colour. This formation occurs most commonly in bronchiectasis, and very typically also in putrid bronchitis, and gangrene of the lungs.

In one of the rarest diseases known to physicians, " plastic bronchitis," peculiar formations termed **" bronchial casts "** or **" stolons "** (fig. 48) are expectorated. The sputum in this disease is either of a catarrhal type, or contains a large proportion of blood. In it are masses covered with mucus rolled up in the form of balls, and stained with blood. If these are thrown into water they spread out, and when freed from mucus are seen to consist of complete casts of parts of the bronchial tree. According to Biermer they may extend to its finest subdivisions, so that the minute terminal filaments may show bulbous ends, due to their having

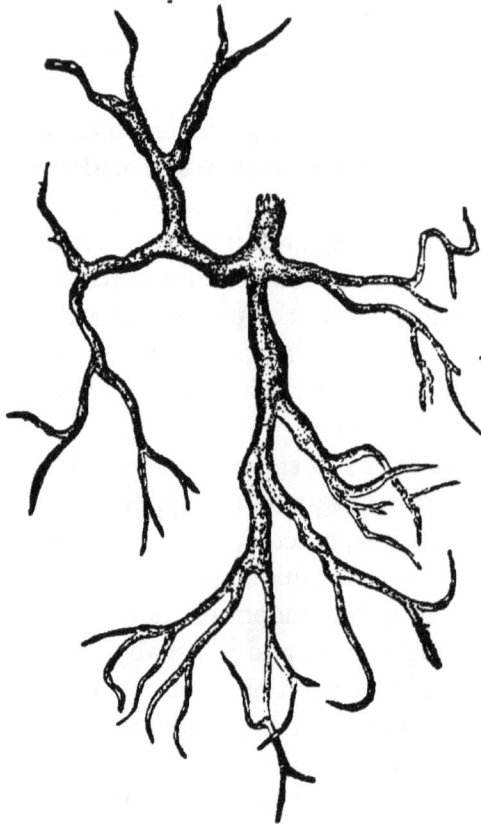

FIG. 48.—Bronchial stolon (Bizzozero and Firket).

been moulded in the infundibula. Their colour is a greyish-white, sometimes, however, having a reddish tinge. This depends either upon their being merely stained with blood externally, or owing to their central cavity being filled with blood clot. In the great majority of cases this central bore contains bubbles of air, producing an appearance as of a string of pearls within them (Fränkel).

When examined more minutely the casts are found to be composed of a number of. concentric laminæ, separated at intervals by narrow spaces, and with a central cavity running their whole length, except in the finest branches.   When submitted to microscopic examination they are found to consist of a fibrillar or hyaline base, in which are embedded large numbers of leucocytes. Charcot-Leyden crystals are often found in quantities within the meshes.   The length of the casts is usually $1\frac{1}{2}$ to 2 inches, but sometimes they are much longer, even attaining 7 inches.   Their diameter is narrow, rarely exceeding that of a goose quill.   They divide dichotomously, the branches gradually diminishing in length and thickness.   A slight bulging is often seen at the points of division, dependent on a similar condition in the branches of the bronchi themselves.   According to Biermer the site of formation of the casts may be inferred from the length of the intervals between successive points of bifurcation.   The short, rapidly branching stolons being derived from the tubes of the upper lobe of the lungs, and the comparatively longer tubes from the lower lobe.   The masses expectorated at different times by the same patient are often so similar in size and arrangement as to give forcible corroboration to the suggestion that they have been produced in the same tubes.

Similar bodies are found occasionally in pneumonic sputum, and in the rare cases in which diphtheritic membrane has penetrated to the finest branches of the bronchi.   The casts in such a case are not so complete as in plastic bronchitis, and are thicker and less elastic.

Casts very similar to the above are occasionally seen after an hæmoptysis, they are easily distinguished from the stolons, being homogeneous in structure, more transparent, and only consisting of a few branches, these being difficult to trace, and to separate from one another.

We need not in this little work enter into fuller details of the characters of sputa diagnostic of the different diseases of the respiratory organs, but will now pass to the

MICROSCOPIC EXAMINATION.

The same precautions must be taken in collecting the samples as were described in preparing for the macroscopic examination, and especially is the sputum to be selected which is coughed up early in the morning, before any food has been taken.

As a rule special portions have to be picked out and placed beneath the microscope, and this process is greatly aided by emptying the spittoon into a black vulcanite dish, similar to those employed by photographers, or into dishes specially prepared for this work. By this means the opaque-white particles which yield the most useful information are rendered conspicuous and more easily recognisable.

These portions are most conveniently removed by forceps and scissors; it is a good plan to slightly bend the tip of one of the blades of the forceps, so as to form a small hook. Another method very generally adopted is to employ two steel pens, these act admirably, both as tearers and lifters.

Some aid is rendered to the general microscopic examination of sputum by pouring it into a large conical glass containing water, the heavier portions, such as fragments of lung tissue, spirals, pieces of membrane, &c., will fall to the bottom after a few hours, and may be removed by means of a pipette.

In the preliminary examination any small opaque portions are removed and placed in the centre of a glass slide, and a cover-glass laid on the specimen, and gentle pressure applied in order to obtain a uniform layer. If it be desired to preserve the specimen a drop of glycerine must be added, then any superfluous moisture removed from the edge of the cover-glass by means of filter paper and liquid balsam run round its edges. Certain objects, such as leucocytes, pus cells, salivary corpuscles, &c., will be found in all sputa, and in others objects of more diagnostic value.

We now proceed to describe in detail the various formed elements which may be met with.

1. **Leucocytes and mucus corpuscles** (fig. 49, ƒ).—In greater or less numbers these are constant ingredients of all sputa. The latter are less granular in appearance than the former, and are possessed of more nuclei ; they are larger too than the leucocytes, but owing to the effects of diffusion it is often almost impossible to distinguish between them. In mucoid expectoration the mucous corpuscles naturally predominate, whilst in purulent sputa the pus cells (leucocytes) may occupy the entire field. Both varieties of cells are seen in all stages of fatty degeneration, and may contain pigment granules, particles of carbon, &c., and after an hæmoptysis, crystals of hæmatoidin. It will be obvious from these remarks that very few deductions can be with safety made from their presence. Closely.

resembling these cells, but rather larger, are the "*sali-vary corpuscles.*"

2. **Compound granule cells** (fig. 49, *e*).—These bodies are found in most sputa, often in so great numbers as to conceal all the other elements.   In general appearance they resemble leucocytes, but are three or four times as large.   In shape they are round or oval.   Their protoplasm is coarsely granular, and they contain several large oval nuclei which are ren- dered more conspicuous by the addition of a drop or two of dilute acetic acid.   They are frequently the seat of particles of carbon and dust, and often appear in an extreme stage of fatty degeneration.

Again, sometimes peculiar formations are found in them, first described by Virchow ("Archiv.," Bd. vi., page 562), and termed by him **"myelin drops"** (fig. 49, *d*), owing to their similarity in appearance to coagu- lated nerve pulp.   These objects are irregular in form, being sometimes circular, sometimes spindle- and some- times club-shaped, and having a tendency to flow to- gether and form unshapely masses.   They are also found floating free in the general mass of mucus.   The chemical nature of these bodies is uncertain, but most authors consider that they consist of lecithin.

The origin of the compound granule cells has been much disputed.   The majority of observers ascribe their source to the alveoli of the lungs.   Against this view is the fact that they are not only found in conditions in which the alveoli are affected, but also in bronchitis and in the morning expectoration of perfectly healthy people. In favour of this view on the other hand, as before stated, is the fact that they are the principal seat of foreign par- ticles which have been inhaled into the lungs, and are especially numerous in miners, &c., suggesting the idea that a slight irritation is set up, sufficient to loosen these cells, and so cause their presence in the pulmonary secretion. Other authors, of which Cohnheim is the chief, consider that they are simple lymph corpuscles which have become enlarged and swollen owing to the imbibition of cellular débris, &c. I most certainly incline to this last view.

3. **Red blood cells** (fig. 49, *g*).—The subject of hæmoptysis has already been fully considered. The red corpuscles, contrary to what occurs in the urine, retain

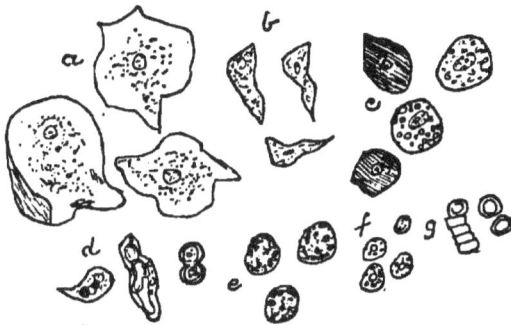

FIG. 49.—Various cells in sputum. *a.* Squamous. *b.* Columnar. *c.* Cubical. *d.* Myeloid. *e.* Compound granule cells. *f.* Leucocytes. *g.* Red blood cells.

as a rule their shape and colour. When they are very numerous they form rouleaux, and preserve their ordinary histological behaviour. Under some circumstances, as in pneumonia, the corpuscles yield up their colour to the general mass of sputum, and then appear as colourless rings, and the same process of destruction is also noticed in gangrene of the lungs. In pneumonia they may only be represented by crystals of hæmatoidin.

4. **Epithelium.**—Three varieties of epithelium are

found in the sputa, namely, squamous cells derived from the buccal cavity, pharynx, and upper part of the larynx; columnar cells from the greater portion of the respiratory passages; and cubical cells from the terminal bronchi and alveolar walls.

Squamous cells (fig. 49, *a*) are always found, and therefore have no pathological interest. When heaped together in big masses their edges are very likely to be mistaken for elastic fibres, but careful focussing will obviate this error.

The *columnar cells* (fig. 49, *b*) are of more value in diagnosis, for when they occur in large numbers they denote an inflammatory state of the parts from which they are derived. They are usually more or less altered from their original form becoming swollen and pear-shaped, and their protoplasm more hyaline in appearance, and through partial loss of their contents form the well known "goblet cells." Although in their original state they are furnished with cilia, these appendages are very seldom preserved in the sputum. The cilia occur under only two conditions, namely, strong mechanical irritation of the bronchial mucous membrane, or ulcerative destruction of the same. From the first cause it is that they are sometimes seen in asthmatic sputum, for here the smaller bronchi are irritated by efforts to expel the thick and tenacious masses of secretion (Fränkel).

Cubical epithelium (fig. 49, *c*) is derived from the alveoli; the cells are not often met with in their original condition, but are more or less altered by fatty degeneration. In the expectoration of patients suffering from brown induration of the lungs, especially in that ejected in the early morning, small reddish-brown streaks and particles are found, which when examined microscopically are found to consist of masses of pigmented epithelial cells, the particles of pigment being circular or irregular, rarely crystalline. According to some authors the compound granule cells which have been just described are considered to be altered alveolar epithelium.

5. **Elastic tissue** (fig. 50).—In the microscopic examination of sputum the search for elastic tissue stands second only in importance to that for tubercle bacilli. This tissue may have its source in the alveoli of the lungs, in the mucous membrane of the respiratory tract, or in the vocal cords and epiglottis. In the large majority of cases, however, it is derived from the alveoli.

When present it is essentially diagnostic of partial destruction of the lungs or other parts of the respiratory tract. A little experience will very soon enable the observer to recognize the fibres derived from the parenchyma of the lung from those of the bronchi or larynx.

In order to search for elastic tissue, the sputum for twenty-four hours should be collected and thrown into a black dish. Sometimes the collections of elastic fibres are so large that they appear as opaque, white specks floating in the mucus. When one of these is removed and placed under the microscope it will be seen to consist entirely of large meshes of elastic tissue, showing quite plainly an alveolar arrangement, roughly tracing the forms of the alveoli. Such cases, however, are not common, and as a rule one must search in the general mass of expectoration for any particularly opaque particles, remove them by means of forceps and scissors, and place them on a slide. A cover-slip is next laid on, and gentle pressure used to obtain a uniform layer. The specimen must then be examined with a ¼ inch lens, and after a little practice the curled fibres will soon be recognized. By this method they retain exactly their original form, and much useful information may be gained from a study of them. If the sputum be very thick and tenacious, some aid in rendering it more transparent may be gained by running under the cover-glass a drop or two of a 30 per cent. solution of caustic potash. This dissolves the mucus and other matters, leaving the elastic fibres untouched. Several preparations should be examined before a negative result is announced.

Another method which is frequently practised is one which was first introduced by Dr. Samuel Fenwick. It has the advantage of being more certain of detecting elastic tissue if present, but the appearance of the fibres is considerably changed, and many characteristics presently to be alluded to are lost. After considerable experience I have come to the conclusion that the former method, although perhaps more tedious, is to be preferred.

Dr. Fenwick's method is carried out as follows:—The sputum is collected for twenty-four hours, and poured into a beaker. An equal quantity of a solution of caustic soda, twenty grains to the ounce is then added, and the mixture boiled until it becomes perfectly liquid, being occasionally stirred with a glass rod. When this stage has been reached the heating must immediately cease, for over-boiling will either dissolve the elastic tissue or render it so transparent and irregular in appearance that it can scarcely be recognized. When the boiled fluid has somewhat cooled it is thrown into a conical glass containing about four times its volume of water. In an hour or so a deposit will have formed in which the elastic fibres will be found, and portions of it may be removed with a pipette and examined under a microscope.

Elastic tissue is composed of fasciculated fibres, highly refractive, but possessing dark contours. They branch dichotomously. They have a slightly yellowish tinge, and have a tendency to break across, leaving a sharply defined edge. They usually occur in wide hoops and curves. Their clearly cut outline is especially characteristic, and enables them to be distinguished

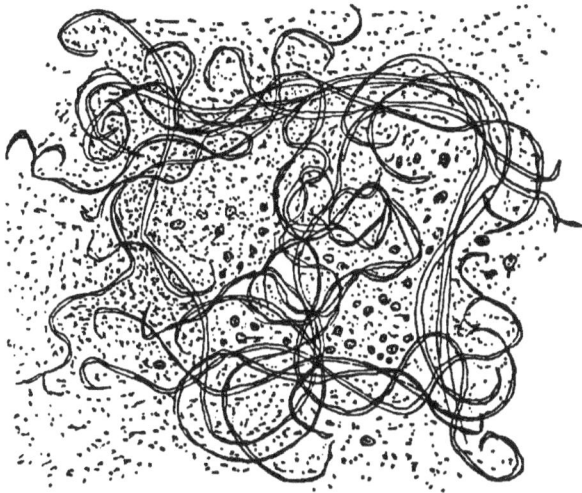

FIG. 50.—Elastic tissue.

from other bodies such as cotton and silk fibres, for which they might otherwise easily be mistaken.

When derived from the parenchyma of the lung they will be seen to have a distinctly alveolar arrangement, as is well shown in the accompanying figure (fig. 50) which was procured from the expectoration of a patient about twelve hours after an injection of tuberculin had been administered.

When coming from the larynx the arrangement of the

fibres is more linear, the fibres being straighter, and devoid of the bold curves which are so characteristic of the last named variety.

Fibres from the other parts of the mucous membrane of the respiratory tract are similar to those coming from the larynx in their general arrangement, but are finer, and the bundles are generally of smaller dimensions.

Their diagnostic value is great. Roughly speaking, their presence indicates that a destructive process is occurring somewhere in the respiratory tract. If from the larynx or bronchi there is probably considerable ulceration. If derived from the former the laryngoscope will give a far more satisfactory diagnosis than the contents of the spittoon ; but it is in those cases in which the arrangement of the fibres denotes that the parenchyma of the lung is affected that their recognition proves of so much value.

It was at one time thought that they were diagnostic of tubercular disease of the lung. Troup ("Sputum ; its Diagnostic and Prognostic Signification ") maintains that where the elastic tissue of the pulmonary alveoli is seen, the correct diagnosis in nine cases out of ten will be that of tubercular phthisis. If tubercle bacilli be also found of course the diagnosis is assured, but the same author considers that careful microscopy of the expectorated matters can frequently detect its presence in a very early stage of phthisis, even when percussion and auscultation practised by skilled observers give only doubtful or negative results. Dr. Troup even goes further than this, and states that he has frequently found elastic tissue months before bacilli made their appearance. I cannot say this has been my experience, and would most certainly give the palm to the tubercle bacilli for early diagnostic worth.

In addition to being found in phthisis, elastic tissue is occasionally seen in the sputum of pneumonia even when the disease has run a normal course (v. Jaksch). The fibres are also found in cases of abscess of the lung, and they are then seen on careful examination to be thickly set with pus corpuscles, and to be greyish-yellow in colour, especially at their edges. They are also found in the sputum from bronchiectatic cavities.

In cases of gangrene of the lung these fibres are again occasionally found. They are generally of a dirty colour with jagged edges, and are surrounded by greyish-yellow masses. Their appearance in this disease is not very common, in fact rather the exception, the reason probably being that though there is a considerable destruction of lung tissue the elastic tissue is dissolved by the ferments evolved during the process. Fränkel gives a somewhat similar explanation, suggesting that the fibres are destroyed by the numerous bacteria which are present.

In addition to the information yielded by the presence of elastic tissue of disintegration of lung tissue, more precise knowledge may be obtained by closer examination, as has been pointed out by Sir Andrew Clark. Thus if complete alveolar rings are present which are at the same time very elastic, we may assume that there is a rapid destruction of lung tissue taking place.

If small "tailed" pieces only are seen which have lost their elasticity, a more chronic process is indicated, whilst occasionally fibres are found encrusted with lime salts (fig. 51) and these must have lain for a considerable period in the lung cavity.

Care must be taken not to mistake other bodies which are often present in the sputum, for elastic

tissue ; such objects are portions of fungi, as the Leptothrix buccalis and various forms of aspergillus. These
are not so regular in outline as the elastic fibres and
are generally surrounded by spores. A description of
these fungi will be given later on.

But the bodies which are most likely to mislead are
fibres of cotton, silk or flax (see p. 278) ; a little practice, however, will soon prevent the observer from
making any such blunders. Fibrillation of the mucin
of viscid sputa will sometimes give rise to an appearance, at first sight, closely resembling a bundle of fibres,

FIG. 51.—Fragments of elastic tissue partially encrusted with
lime-cells.

but careful focusing will easily discern between the
two.

6. **Fragments of connective tissue.** — These
only occur in rare cases and are derived from the
walls of the alveoli and bronchi. They occur as dark
grey particles, and are easily recognised. They are
chiefly found in cases of gangrene and abscess of the
lung. In ulcerative processes affecting the larynx, small
portions of cartilage may be loosened by the violence of
the cough, and be ejected with the expectoration.

7. **Curschmann's spirals** (fig. 52).—These peculiar bodies are chiefly found in the sputa of asthmatic
patients, where they are sometimes exceedingly nume-

rous. They were at one time considered characteristic of asthma, but have since been shown to occur, although only in small numbers, in the expectoration of pneumonia and of acute and chronic bronchitis. They were first minutely described by Curschmann, and hence their name.

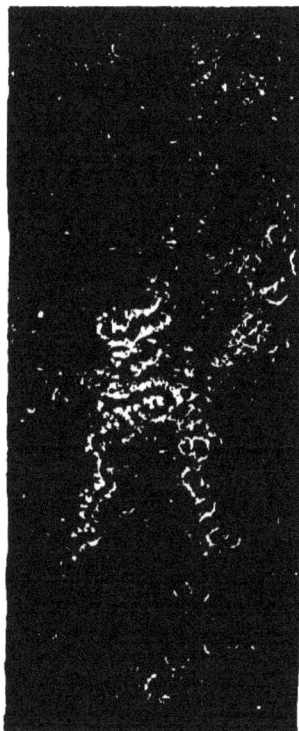

FIG. 52.—Curschmann's spirals (Troup).

If the sputum from a case of bronchial asthma be poured out on a black plate small opaque white particles will be seen which, even with a low power lens, may be seen to consist of spiral formations. When one of them is placed under the microscope, an intricate arrangement of spiral threads can be discerned. In the centre there is a highly refracting thread sometimes straight and sometimes twisted corkscrew fashion. Around this, in a complicated manner, is entwined an ensheathing layer of fine threads, in the meshes of which are seen leucocytes and epithelial cells.

Some of the spirals will be noticed to possess a peculiar prominence, and if these be more closely examined a large collection of pointed octahedral crystals of various sizes will be found.

The exact structure of these spirals has not yet been definitely settled. According to some observers they consist largely of collections of bacteria, but the majority of authors consider that they are formed of mucus which has been moulded into spiral shape in its progress through the small bronchioles. The central core is due partly to an optical effect and partly to the greater compression which it has apparently undergone.

Three modifications of the spirals are generally met with. Firstly, the complete spiral, consisting of the central thread surrounded by the mass of mucoid material; secondly, the central thread alone is often seen, and thirdly, the spirally twisted portion of mucus may occur without the central thread being visible. The diagnostic value of these bodies has not been established with any certainty. They are probably due to a catarrh of the finest bronchioles, and as before stated they are most commonly associated with bronchial asthma.

8. **Tonsillar casts** (fig. 53).—Casts of the crypts of the tonsil, or more frequently fragments of such casts are occasionally found in the sputum.

They chiefly occur in connection with acute inflammatory diseases of the tonsils, but in comparison with the clinical aspects of the case have little diagnostic value. They bear considerable resemblance to masses of mucus, and are consequently easily overlooked, but they are more likely to be mistaken for small leashes of elastic tissue. Tonsillar casts appear as irregularly-shaped bodies composed of parallel fibres finely knit together into a close net-work. When teased out the

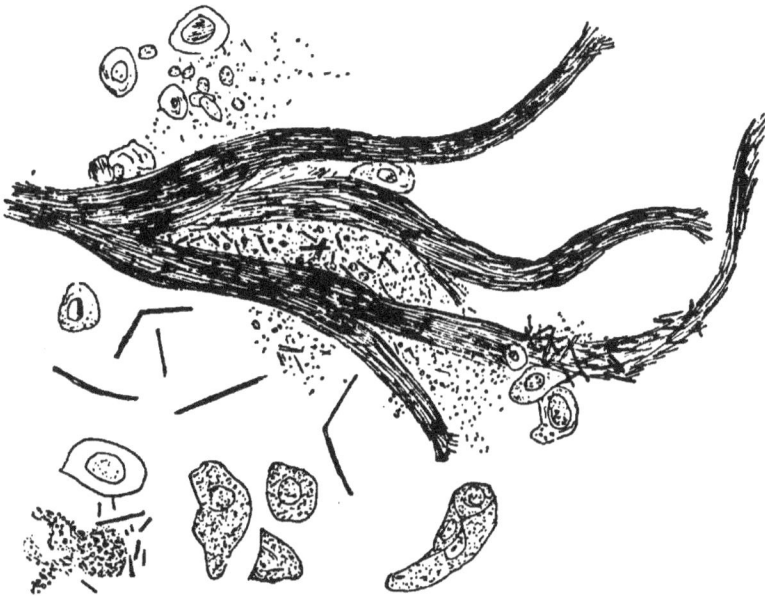

FIG. 53.—Cast of a tonsillar crypt (from a drawing kindly lent by Dr. Sheridan Délapine).

appearance represented in the figure is produced. A complete cast assuming the exact form of a tonsillar crypt is rarely met with, but as already stated, more commonly only fragments are found.

9. **Corpora amylacea.**—These bodies are very rarely met with; they were first described by Friedreich in 1856. They appear as oval or circular grains and

are very similar in form to starch grains. They are composed of a vitreous looking substance, stratified in irregular layers around a central nucleus, which differs in appearance from the peripheral layers which are arranged so as to present a radiated formation. The central portion is sometimes scooped out so as to form a small hollow from which shallow fissures extend, and occasionally is filled up with a material of different composition to the rest of the grain.

These amyloid bodies are characterised by their chemical reactions. If a solution of iodine in iodide of potassium be run in under the cover-glass a deep blue colour is produced, which becomes almost black on the addition of sulphuric acid ; the nucleus, however, is stained yellow by the iodine. Methyl violet causes the same reaction as in amyloid disease generally (see p. 101), colouring the grains red, whilst the surrounding tissues assume a blue tinge. Picro-carmine has a peculiarly selective action, staining the bodies a deep red. With methylene blue the corpora amylacea assume an emerald-green colour.

As regards the possible significance of these amyloid concretions many diverse theories have been suggested, the two principal being those of Friedreich and Langhans. The former considers that they are derived from remains of hæmorrhoids in the lungs. In all his cases there was a history either of recent hæmorrhagic infiltration, or traces of past inflammation. His theory is that a minute clot forms locally, thus constituting the nucleus, and fibrin is deposited in layers around it, and that this formation undergoes modification (of what kind is at present unknown) which causes the bodies to assume their characteristic appearance and properties.

Langerhans on the other hand assigns to the corpora amylacea a cellular origin. In this view he has but few supporters, and moreover the bodies described by him do not agree in all particulars with those now under consideration. The observations of Jurgens and Zahn, and those more recently of Curtis, corroborate those of Friedreich.

10. **Portions of membrane.**

a. **Diphtheritic membrane.**—In cases of membranous laryngitis, whether occurring after scarlet fever or being undoubted instances of diphtheria, portions of the membrane are often coughed up and should be submitted to careful microscopic examination.

The membrane consists of a soft, thin, whitish layer presenting fine striation and wavy lines, in the meshes of which are pus cells and epithelial débris. The fibrillated substance and cells alternate so as to produce a laminated structure.

To demonstrate its characters it should be placed in water when it can easily be spread out.

Portions of it should be stained in methylene blue (see page 91) in order to ascertain whether any organisms be present. In the membranous laryngitis occurring after scarlet fever this is never the case.

In diphtheria (it being assumed that "croup" and diphtheria are the same disease) the pathogenic organisms are often found, though not invariably so. They consist of bacilli of various shapes, the organism being polymorphous.

b. **Echinococcus membrane.**—Hydatid disease of the lung is an extremely rare one ; there are, however, several cases on record. The membranes are the same as those found in disease of the liver, and further description will be postponed until a future chapter.

T

Occasionally the cysts themselves or loose hooklets are expectorated, rendering the diagnosis easy.

The source of these bodies, however, is not necessarily the lungs, as they may arise from a suppurating hydatid cyst of the liver perforating through the diaphragm into the lung.

11. **Crystals.**—Many forms of crystals are found in the expectoration, but few of them have any diagnostic importance.

a. **Hæmatoidin.**—These crystals occur in the form of ruby-red rhombic prisms or needles, often grouped together into bundles. Complete crystals are not often met with in the sputum, hæmatoidin more commonly being found as irregular red particles. Both crystals and fragments are frequently enclosed in white blood corpuscles.

Their presence indicates that blood has lain for some time in the air passages, or that an abscess has perforated into the lungs. They often occur in large quantities after an hæmoptysis, or some time after a pulmonary infarct.

If hæmatoidin is found chiefly enclosed in cells, an old hæmorrhage of the lungs is the most probable cause, but if the crystals are chiefly free there has most likely been a rupture of an abscess from some neighbouring organ into the lung.

b. **Charcot-Leyden crystals** (fig. 54).—In the sputum of asthmatic patients these crystals are often found in large numbers. They are long and pointed octahedra, or sharp or truncated fusiform bodies; they vary greatly in size. They are soluble in alkalies, mineral acids, and warm water, but are insoluble in ether, alcohol and cold water. Their exact chemical composition is not known. Schreiner supposes them to consist of phosphoric acid combined with an organic base.   Curschmann and Ungar are of opinion that these crystals are derived from the decomposition of organic matter in the bronchioles.   In connection with this, v. Jaksch's remarks are worthy of note ; he says that in fresh cases of bronchial asthma, or at the commence-ment of a new series of attacks, Curschmann's spirals are chiefly found, but no crystals ; if such prepara-tions are prevented from evaporating for about forty-eight hours, Charcot-Leyden crystals will be found to

FIG. 54.—Charcot-Leyden crystals.

have been formed.   In the further course of the attack the sputum when first ejected will contain the crystals. This seems to suggest that the crystals are formed by the decomposition of the spirals.

They are not absolutely diagnostic of asthma, being present in other conditions.

c. **Cholesterin crystals** (fig. 43, p. 210).—The well known rhombic, iridescent plates, with notches at one corner, are not often met with in sputum.

The expectoration of phthisis, or the pus from a pulmonary abscess or empyema, are the most frequent conditions under which they occur. They generally lie together in heaps, are easily soluble in ether, but insoluble in acids, alkalies and water. When treated with dilute sulphuric acid and tincture of iodine, the crystals gradually change their colour, becoming violet-blue, green and red. With sulphuric acid alone they slowly turn from yellow to violet-red.

d. **Crystals of the fatty acids.**—These occur as groups of long, sharply-pointed needles, often arranged in small rosettes. They are characterised by their easy solubility in ether.

They are most usually found in connection with putrid bronchitis and gangrene of the lungs, but also occur in the sputa of bronchiectasis and phthisis. They have therefore but little diagnostic value.

As regards their chemical nature, they consist chiefly of palmitic and stearic acids, in combination with sodium, potassium, calcium and magnesium.

e. **Leucin and tyrosin** (figs. 40 and 39).—These two substances are but rare constituents of the sputum. When present they in no way differ in form from that when appearing in urinary sediments.

They have been described by Leyden in a case of putrid bronchitis, and in another in which an empyema burst into the lung.

f. **Oxalate of lime** (fig. 39, p. 199).—These occur occasionally in their characteristic envelope-shaped crystals, or as an amorphous sediment. They have no diagnostic significance.

g. **Triple phosphates** (fig. 41).—In appearance these crystals are the same as those described on p. 206. Being soluble in acids, they are only found when the sputum is alkaline. They sometimes occur in the sputa when a sanious exudation has broken into the lungs, and when decomposition of albuminous matters has taken place with evolution of free ammonia.

12. **Concretions.**—It not uncommonly happens that small portions of cretaceous matter are coughed up by phthisical patients. When these particles are of a large size, they are known as "lung stones." Fränkel also describes "bronchial stones." These latter show indications of branching, and are of the size and shape which would be expected if they had been formed in the bronchi.

The origin of both varieties is uncertain. Andral describes a case in which after death in a patient forty years old, who died of phthisis, he found a copious deposit of chalk in the cartilaginous portions of the smaller bronchi; the mucous lining was in some places ulcerated, and some of the calcareous matter had escaped. Another source suggested is calcification of caseous matter which has been retained in a pulmonary cavity. Finally, some observers consider that these cretaceous particles are derived from bronchial glands which have become calcified and ulcerated through into the bronchi. As regards their size, this varies from that of a small pea to that of a plum-stone, or even of a walnut. The number ejected by one patient may be very great. Andral records a case of a phthisical patient who, in eight months, expectorated no less than 200 stones, the largest of which was the size of a cherry.

13. **Foreign bodies.**—For the most part these have little diagnostic worth. Some of them have evidently been inhaled, and again ejected, sometimes in so changed a condition as to be hardly recognisable.

Such are particles of food which occasionally lie for a considerable period in the ventricles of Morgagni.

Other instances of foreign bodies so expectorated are nut shells, teeth, threads, &c. The retention of these

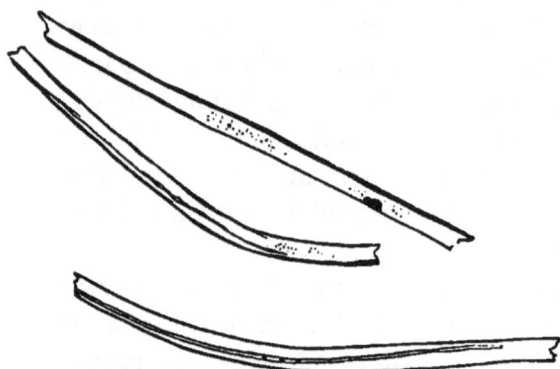

FIG. 55.—Silk fibres.

bodies in the air-passages often gives rise to seriou symptoms, the explanation of which is only afforded on the expulsion of the offending body.

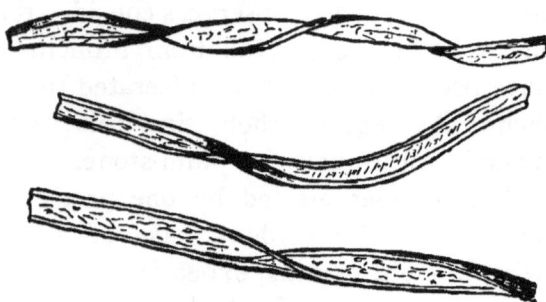

FIG. 56.—Cotton fibres.

Occasionally when malignant disease of the larynx is present, portions of tissue slough off and are ejected. Microscopic examination of such fragments may yield most important aid in diagnosis and prognosis.

Dr. Fagge records two cases in which hairs were expectorated from mediastinal dermoid cysts.

FIG. 57.—Linen fibres.

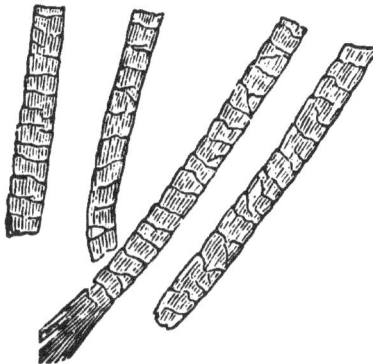

FIG. 58.—Wool fibres.

14. **Accidental ingredients.**—In the. opening part of this chapter stress was laid on the necessity of the prevention of extraneous matters from entering the

spittoon. But sometimes in spite of all care such accidental substances will find their way into, and become
mixed with the sputum. Most commonly these are
particles of food matter, such as animal and vegetable
tissues, globules of fat, starch grains, &c. Bundles of
elastic tissue derived from the food may be distinguished from those coming from the patient by being
coarser in texture, and arranged in a more parallel
manner. Starch grains may be known from the corpora amylacea by their chemical reactions, the former
assume a purple colour when treated with iodine dissolved in iodide of potassium.

Fibres of silk, cotton, linen, &c., are also often met
with.

Fibres of silk (fig. 55) are tapering, possess sharp
outlines, and do not branch. Fibres of cotton (fig. 56)
are twisted, and have curious central markings. Fibres
derived from linen (fig. 57) are jointed, cylindrical,
have ragged uneven ends, and fine branching filaments
at intervals. The fibres of wool (fig. 58) are rounded,
with fine cross-markings and reticulations in the border
at the site of the cross-markings. The central longitudinal canal is usually invisible. Any of these fibres
may be coloured.

To the above list must be added hairs, particles of
dust, chips of wood from the floor, &c.

Many of the above objects, unless carefully examined,
may give rise to much misapprehension.

CHAPTER XVII.

THE MICRO-ORGANISMS OF SPUTUM.

PROBABLY no department of medicine has made greater advances during the last few years than bacteriology, and this has been largely made use of for clinical purposes in the examination of sputum.

In this chapter will be described the most convenient and reliable methods employed for the detection of micro-organisms in the expectoration.

1. **Tubercle bacilli.**—A great number of processes have been from time to time introduced for the clinical detection of the bacillus tuberculosis in sputum.

The method which long experience has proved to be speedy and reliable will be here described at length, short reference only being made to other processes.

At the risk of some repetition of the directions given in the previous chapter, this method will be detailed throughout.

In the first place the patient must be directed to reserve that portion of expectoration which has been ejected in the early morning before any food or drink has been taken. If nourishment has been partaken of during the night, it is well to request the patient to carefully rinse out his mouth with some water before using the clean spittoon. Nothing is more annoying than to have food matter mixed with the sputum. I have received samples so contaminated in this way that a satisfactory examination was out of the question.

The collected sample is thrown out into a dish (preferably a dark coloured one). Search is made for the
small, yellowish, opaque, round particles so common in
phthisical sputa, and if any are found they should be
transferred to a cover-glass. Failing these, any other
opaque portions must be selected. These having been
placed on a cover-glass, another glass is laid on it, and
the specimen of sputum gently spread out by means of
pressure. A layer as thin as possible should be obtained, any excess that exudes from between the glasses
being wiped away with a soft cloth.

The cover-glasses are now separated by a sliding
movement, that is, *not* "sprung" apart. Two films of
sputum are thus obtained, which are allowed to dry in
the air, this process being expedited, if desired, by
holding the glasses between the finger and thumb over
a spirit lamp, and as long as the heat does not burn the
hand it will not injure the specimens.

When dry the film must be fixed to the glass, this is
accomplished by seizing the glasses one by one with
forceps, and rapidly passing them three times through
the flame of a spirit lamp or Bunsen burner, thus co-
agulating the albumen.

A pause must now be made in the description of the
process, to give the composition of the various fluids
required. They are as follows :—

Solution 1.—Neelsen's solution (Ziehl).
Fuschine . . . . 1 part
Dissolved in a 5 per cent. watery so-⎱ 100 parts.
lution of carbolic acid . . . . .⎰
Alcohol . . . . 10 „

Solution 2.—Dilute watery solution of sulphuric acid,
25 per cent.

Solution 3.—Methylene blue.

Methylene blue . . . .   2 parts.

Alcohol 15   „

Water 85   „

To resume, a small quantity of Neelsen's solution is
placed in a watch glass, and heated until steam begins
to rise. Into this the cover-glasses, prepared as previously described, are placed, film-side downwards. It
is sometimes difficult to decide, especially in the later
stages of the process, which is the prepared side of the
cover-glass, but this can be determined either by lightly
scratching each surface with a needle, and observing
whether a mark is made, or more simply still by holding
the cover-glass towards the light, when the side on
which the film is appears dull, whilst the unused side
yields a bright reflex. The specimens are allowed to
remain in the heated stain for *two minutes*. Most books
state that five minutes are necessary, but experience
has proved that two minutes are amply sufficient.

In order to prevent the glasses sticking together,
which they will do when both are submerged, the first
glass should be placed slantingly in the fluid, so that it
sinks. The second glass, however, should be held between the finger and thumb, and dropped on to the
fluid. If this be done from a short distance, the glass
will float.

After a stay of two minutes in the stain the glasses
are removed one by one by means of forceps, and
thoroughly decolourised in the dilute sulphuric acid contained in a shallow capsule. A few seconds usually
suffice for this, although contrary to what is generally
believed, the bacilli retain the stain for at least thirty
minutes in the acid ; they are in fact difficult to decolourise, and it is important that the red stain should be

removed from other bodies as far as possible, so that any excess should be on the side of a longer stay in the acid rather than the reverse. The best mode of procedure is as follows. The glass is retained in the forceps, and gently moved about in the acid, and after a few seconds is transferred to a dish of water. The colour will probably be partially restored, and the specimen must then be returned for a few seconds to the acid, again rinsed in water, this process being repeated until no more colour is visible.

The specimen is next placed, prepared side downwards, in the methylene blue solution contained in a watch glass. It is allowed to remain for about half a minute, and is again washed in water. It is then allowed to dry, either by exposing it to the heat of the flame of the spirit lamp, or by gently pressing it between two folds of filter paper. It is finally mounted in water if only required for diagnostic purposes, or in balsam if a permanent preparation is desired.

Thus prepared the bacilli (Frontispiece, fig. 1) appear as short red rods, whilst the mass of the sputum is stained blue.

The tubercle bacilli are of variable size, but have an average length of half the diameter of a red blood corpuscle. They are usually straight, but may be found slightly curved. Their grouping is very characteristic, being generally arranged in small numbers of two or three, lying cross-wise or at angles to one another. Sometimes, however, they are so numerous as to form large clumps, in which the individual members can only be recognised with difficulty.

Their characteristic property is that of retaining the stain in spite of exposure to acid. By this they are distinguished from all other bacilli, except that of leprosy, which in this country is hardly likely to be met with, and certainly never in expectoration.

Their presence in sputum, therefore, is absolutely diagnostic of a tubercular process seated somewhere in the respiratory tract, the exact locality being soon recognisable by the stethoscope or laryngoscope.

A peculiar beaded appearance is often noted, supposed by some to be due to the presence of spores which will not retain the stain. As regards their number, opinions differ considerably whether any definite conclusion as to the extent and activity of the disease can be drawn from observing whether the bacilli are numerous or not. Dr. Theodore Williams considers that a large number of rods denotes rapid spread of the disease, but I am decidedly inclined to agree with Dr. Kidd and Mr. Taylor (" Transactions of the Royal Medico-Chirurgical Society," vol. 71, p. 173), who after an exhaustive examination of the results at the Brompton Consumption Hospital, came to the conclusion that no importance, as regards prognosis, could be attached to the large or small number of rods seen in the specimens.

A negative result is of little value and proves nothing, but if after repeated examinations of several satisfactory samples of sputum no bacilli be found, this may be taken as presumptive evidence that the case is not a tubercular one, although it is no absolute proof.

A positive result on the other hand is everything, and cannot be ignored. Although no other bacilli stain like the tubercle, yet errors sometimes arise owing to foreign bodies retaining the colour. Such matters as hair and horn are difficult to decolourise, and if the specimen has not remained sufficiently long in the acid, the edges of cells often appear red, but this mistake may easily be remedied by careful focusing.

In order to examine satisfactorily for bacilli, an
Abbé's condenser should be used, and at the lowest a one-
sixth inch lens, preferably a one-twelfth, although they
may be seen with a quarter. The method given above
for staining tubercle bacilli can be thoroughly relied
upon, but there are several modifications of it and other
methods, only a few of which will be considered here.

A rather more rapid process may be conducted with
the same stains.

The cover-glasses are prepared in precisely the same
way, but instead of being placed in a watch glass, a
couple of drops of Neelsen's solution are placed on the
prepared side of the cover-glass, which is seized in
forceps. The glass is held over the flame of the spirit
lamp until bubbles are observed to form ; the excess of
stain is then poured off, and the glass swilled in the
dilute sulphuric acid and washed in water. A drop or
two of the methylene blue solution is then placed on the
specimen, allowed to remain for half a minute, and
washed off in water. The preparation is next dried,
and examined in balsam or water.

According to Ehrlich it is very difficult to obtain the
acid solution without free acid fumes, which tend to
decolourise the bacilli as well as the mucus. He therefore makes use of the following process :—

The staining solution must always be prepared fresh.
A few drops of pure aniline oil are placed in a medium
sized test-tube which is filled with water, the mixture
being thoroughly shaken and filtered through a double
fold of paper. A little of the clear fluid is placed in a
watch glass, and a few drops of a saturated alcoholic
solution of fuchsine are added until the solution just
becomes opaque. This being most easily judged by
placing the watch glass on the edge of a piece of filter

paper, and observing when this edge becomes obscured. Cover-glasses are floated, prepared side downwards, on this fluid for several hours (six to twelve). They are decolourised in a mixture of one part of nitric acid to two parts of a saturated solution of sulphanilic acid. After being swilled in water and rendered quite colourless by re-immersion in the acid solution, if necessary, they are counter-stained with methylene blue, washed in water, dried, and mounted in balsam. The rationale of this process is that the sulphanilic acid absorbs the nitrous fumes, so allowing the pure acid to act on the bacilli.

The process recently introduced by Dr. Gabbet seems to be an excellent one. It closely resembles those already described, but is more rapid; on the other hand one has not the same control over the process of decolourisation.

The first staining solution has the following composition :—

Carbolic acid (5 per cent.) . . . . 100 c.c.
Magenta . . . . 1 gramme
Spirit . . . . 10 c.c.

Some of this solution is placed in a watch glass, the cover-glasses immersed in it and heated until steam begins to arise. They are then placed for one or two minutes in a solution, which at once decolourises and counterstains them. This solution is made as follows:—

Sulphuric acid (25 per cent.) . . . 100 c.c.
Methylene blue . . . . 2 grammes.

On being taken out the specimens are washed in water, and mounted in the usual way.

In some cases in which no bacilli are found by the above methods, but where the history and physical signs point to a tubercular process in the respiratory tract, Biedert's method ("Centralblatt für Klinische Medicin," 1891, No. 6) should be adopted.

To a tablespoonful of the sputum seven or eight drops of caustic soda are added, and two tablespoonfuls of water. This mixture is boiled until it becomes liquid. Four to six tablespoonfuls of water are now added, and the mixture again boiled until the whole is a thin uniform liquid. It is next thrown into a conical glass and allowed to stand from twenty-four to forty-eight hours. Part of the sediment which results is removed with a pipette to cover-glasses, spread out in as uniform a layer as possible, dried, and then stained for tubercle bacilli by one of the methods already detailed.

The only other process for the detection of tubercle bacilli that will be described here is the one introduced by Dr. Heneage Gibbes. It is very simple and rapid, but in the opinion of the author not so reliable as the foregoing. Dr. Gibbes has a special staining fluid prepared as follows :—

Rose aniline hydrochlorate . . . . . 2 parts
Methylene blue . . . . 1 part.

The mixture is triturated and three parts of aniline oil dissolved in fifteen parts of rectified spirit is slowly added, and finally fifteen parts of distilled water.

Cover-glasses are prepared in the usual way, and immersed in the stain, which has been heated until steam has arisen, for five minutes. They are next washed in methylated spirit until no more colour can be extracted, when they are dried and mounted.

2. **Pneumococci** (fig. 59).—In 1883 C. Friedländer announced that he had found the pathogenic organism of pneumonia. In the rusty sputum of pneumonia certain organisms are almost always found, sometimes in very large numbers. Unfortunately, however, for their

diagnostic value, micro-organisms, morphologically precisely similar to them, are found in many other diseases,
and indeed in the saliva of healthy individuals.

These micrococci take the form of spindle-shaped
bodies united at their flat bases, so forming diplococci.
Sometimes three or four are joined together. But what
especially characterises them is the capsule with which
they are surrounded ; they thus appear as oval bodies
provided with a sheath of similar shape.

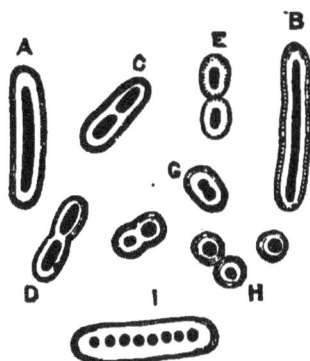

FIG. 59.—Bacterium pneumonia crouposa, from pleural cavity of a
mouse × 1500. *A*, *B*. Thread-forms. *C*, *D*, *E*. Short rod-forms.
G. Diplococci. *H*. Cocci. *I*. Streptococci. (After Zopf, from
Crookshank).

There are two principal methods by which they are
stained.

Cover-glass preparations of the expectoration are prepared in the same way as was described when treating
of tubercle bacilli.

A few drops of aniline oil are placed in a test-tube,
which is filled up with water and thoroughly shaken,
the emulsion being filtered through a double fold of
filter paper. A few drops of a saturated alcoholic solution of gentian-violet are added until the mixture
becomes opaque. The cover-glasses are immersed in

U

this solution for five minutes, and are then swilled for a few seconds in absolute alcohol, washed in water, and mounted in the usual way.

Very satisfactory specimens may also be prepared as follows :—

The aniline water is made as above, but instead of the violet a few drops of a concentrated alcoholic solution of fuchsine are added. The cover-glasses are immersed in this stain for about a minute, and are then placed for two minutes in a watery solution of methylene blue. Finally, they are washed in water, dried, and mounted in balsam. Stained in this way the diplococci are coloured blue, the capsule assuming a rose tint, whilst the ground substance is a bluish-red.

A. Fränkel found a coccus precisely similar in appearance to the above, not only in the sputa of pneumonia, but also in other diseases, and even in normal saliva. I have found very characteristic specimens in the sputa of phthisical patients. It is conceivable that these organisms are normally present in saliva and only excite pneumonia when other causes combine to produce a favourable pabulum upon which they may grow, and evolve the poisons which are necessary to cause the conditions found in the disease. Fränkel's coccus and its capsule stain readily in methylene blue. After cover-glasses have been prepared in the usual way they are immersed in methylene blue solution for five minutes and then washed in water acidulated with acetic acid.

3. **Actinomyces.**—This disease has been very rarely diagnosed in this country during life. Dr. Douglas Powell read a paper before the Royal Medico-Chirurgical Society (see "Transactions," vol. 72, p. 175) on a case which occurred at the Brompton Consumption Hospital.

Very seldom, however, in any country has the characteristic fungus been found in the sputum. Baumgarten,
Paltauf and others, have recorded such cases. Possibly
if the sputum from obscure cases was more frequently.
examined, more cases could be added to the list. The
expectoration in actinomycosis is usually small in
amount, muco-purulent, often of putrid odour, occasionally stained with blood, and contains minute white
or yellowish granules about the size of a small pin's
head. When one of these granules is examined with a
quarter-inch lens, a number of small oval bodies (clubs)
are seen surrounding an indistinct mass, but to bring
out clearly the appearance of the fungus, specimens
must be stained in the following manner. A few of the
granules are placed on a cover-glass, allowed to dry,
and passed three times through the flame. It is stained
by Gram's method (see p. 95), and afterwards counter-
stained by immersion for twenty-four hours in a saturated watery solution of safranin, containing two per
cent. of aniline oil, then cleared in xylol, dr.ed, and
mounted in balsam. The clubs of the fungus are thus
stained yellowish-red, and the threads blue. Instead of
the safranin, alcohol containing a little picric acid may
be added after the Gram's process is complete, a stay
of five minutes being necessary; the clubs are then
stained a bright yellow colour.

4. **Micrococcus tetragonus.**—This peculiar organism, as its name suggests, consists of four cocci
united sarcina-like, and surrounded by a hyaline capsule. The cocci are stained by all the aniline dyes,
and are frequently seen in sputum when this has been
stained by methylene blue. The capsule, however, does
not stain.

This microbe is not pathogenic, being found in the

sputa of various diseases, or even in the expectoration
of healthy people. It is most frequently seen in phthisical sputum. According to Koch, it assists in the
destructive process taking place in the lungs.

Gaffney showed that when injected into guinea pigs
and white mice, the animals died of septicæmia.

- 5. **Septic micro-organisms.**—Various forms of
micro-organisms, both rods and cocci, are found in all
sputa, being especially numerous in the expectoration of
bronchiectasis and gangrene of the lungs. They vary
in size, and take the forms of staphylococci, streptococci, &c.

Diplococci are especially common, and are usually
surrounded by a thin colourless zone.

Occasionally zooglœa masses occur, made up of large
numbers of micrococci. These are especially common
in the sputum of phthisis.

6. **Fungi.**—Several forms of fungous growth occur
in the mouth, and are consequently not unusually found
in the sputum. More rarely they grow in the deeper
respiratory passages or even in the lungs themselves.

The most common is the **Oidium albicans** or
thrush (fig. 60), it occurs in the form of branching
threads, composed of elongated cells, amongst which
numerous spores are. seen. They stain with methylene
blue. The **Leptothrix buccalis** (Frontispiece, fig. 2)
often grows round the teeth, and portions being detached and appearing in the sputum may be a source
of difficulty as regards recognition, unless the observer
be familiar with their appearance. Its reaction with
iodine is characteristic. If a little of the tincture of
iodine, or of a solution of iodine in iodide of potassium
be added to the specimen, the threads of the fungus
assume a deep violet colour, which, however, only lasts
a few hours.

The **Aspergillus fumigatus** has also been described by Virchow as occasionally growing in the lung.
In some cases of extensive destructive ulceration in
the lung, especially in pulmonary gangrene, bodies
known as **" Sarcinæ pulmonis,"** are found in the
sputum. They were first described by Virchow and
Friedreich ("Virchow's Archiv.," ci., 401, 1856, and
390, 1864). In appearance they closely resemble the
Sarcinæ ventriculi, but are rather smaller.

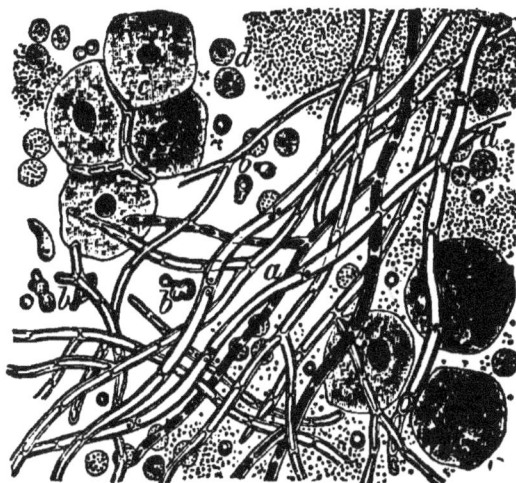

FIG. 60.—Oïdium albicans (v. Jaksch).

The only case in which I have found these bodies,
was one of phthisis running a slow course, the physical
signs being obscure, but tubercle bacilli were present
in the sputum in large numbers.

7. **Infusoria.**—These have been described by Kan-
nenberg ("Virchow's Archiv," lxxv., 471, 1879) in the
sputum of patients afflicted with pulmonary gangrene.
They were chiefly found enclosed in small yellow droplets
surrounded by fatty needles. They were of the monad

and cerco-monad varieties. They exhibited sluggish movements.

According to v. Jaksch the best method of demonstrating them is as follows. A few of the droplets are picked out and placed upon a cover-glass. About three or four drops of a one per cent. solution of common salt are added. The preparation is dried in the air and stained in a watery solution of methyl violet, washed in water, and while still wet placed in a concentrated solution of acetate of potash. By these means the protoplasm of the monads is coloured a beautiful blue.

8. **Entozoa.**—Members of this group are very seldom found in the sputum. In the rare instances of hydatids in the lungs, the hooklets or even vesicles may be discovered.

The Ascaris lumbricoides has occasionally found its way into the œsophagus from the intestines, and has been ejected, and Manson has recorded cases in which the eggs of Bilharzia hæmatobia have appeared, having escaped from the parent worm in the blood vessels.

CHAPTER XVIII.

EXAMINATION OF VOMIT.

An investigation by the microscope of matters which have been vomited, does not yield such useful information as a similar examination of sputum, or even of the fæces. But occasionally, considerable aid to diagnosis is rendered by such means and the student ought to make himself familiar with the appearances which are met with when food has been partially digested. For this purpose he should soak pieces of bread, meat, tendon, cartilage, and the more common vegetables (such as potatoes, cabbages, &c.) in slightly acidulated water, and then tease out particles of these substances on a glass slide, and examine them with the microscope.

Much more valuable information is gained from a chemical examination especially as regards the detection of poisons, but the nature of this work forbids me entering into this subject here.

The vomited matter is placed in a conical glass and allowed to stand for a few hours, when portions of the sediment should be removed with a pipette, placed on glass slides, protected with cover-glasses, and investigated with an inch and quarter-inch lenses. Separate portions of the supernatant liquid should also be examined. The following particles of **food matter** will probably be seen :—

1. Muscle fibres (fig. 61, *b*) recognised by the transverse striations.

2. Elastic fibres, characterised by their clear outlines, and tendency to form curves.

3. Fat globules and needles. The nature of these bodies may be proved by their solubility in ether.

4. Vegetable cells of various forms.

5. Starch granules (fig. 61, *c*) more or less altered by the process of digestion, but stained blue by the action of iodine dissolved in iodide of potassium, a few drops of which should be run under the cover-glass.

Epithelium (fig. 61, *a*).—Two varieties of epithelium may be noticed. Squamous cells occur in small num-

FIG. 61.—*a*. Epithelial cells.   *b*. Muscle fibre.   *c*. Starch granules. *d*. Sarcinæ.   *e*. Yeast cells.   *f*. Non-pathogenic micro-organisms.

bers only, and are derived from the mouth and other parts of the œsophagus, having been swallowed with the saliva.

Columnar cells are sometimes very numerous and then are indicative of catarrh of the gastric mucous membrane. They are generally more or less altered in shape by the action of the secretions. In the vomit of cholera numerous small white flocculi are found, which may be shown to consist of collections of epithelial cells, both squamous and columnar. The occurrence of so-called "cancer cells" in cases of carcinoma of the stomach is doubtful. No doubt cells derived from tumours do make their appearance in the vomit when such conditions exist, but they cannot with certainty be distinguished from squamous epithelial cells. If large irregularly-shaped cells are found in large numbers, together with red blood corpuscles and pigment, this would be presumptive evidence that cancer of the stomach did exist, although without accompanying clinical signs, which are far more constant and valuable, such a diagnosis could not be ventured on.

Red blood cells.—The appearance of red blood cells in the vomit depends upon how long they have been retained in the stomach. If a vessel has ruptured and a considerable quantity of blood has been poured out, they will present their normal colour and biconcave form. As a rule, they are more or less altered by the action of the gastric juice, and are then seen as colourless rings. If the blood has been some time in the stomach the hæmoglobin becomes reduced to hæmatin; a dark brown fluid then results, in which masses of pigment may be seen, recognisable as blood only by means of the spectroscope.

Leucocytes.—A few scattered white blood cells are always present. If they occur in large numbers there is probably gastric catarrh, but the cells may be derived from the saliva, sputum, or nasal mucus which has been swallowed. These cells are seldom present in such quantity that the term pus may be applied, and practically this only happens when an abscess, situated in neighbouring parts, has burst into the stomach or œsophagus. The cells themselves are generally so much altered by the secretions that only their nuclei are visible.

In cases of diphtheria portions of **membrane** may become detached and ejected by the effort of vomiting·

Their nature may be demonstrated by floating them in water, when they will easily spread out.

Vegetable parasites.—Various kinds of fungi are occasionally found in the vomit, some such as leptothrix have evidently been swallowed, and only two are of any importance.

Sarcinæ (fig. 61, *d*).—These bodies appear as round cells grouped together in fours, and very much resemble wool-packs in form. They are of dark silver-grey colour, and are stained mahogany-brown with a solution of iodine in iodide of potassium. They are generally associated with obstructive disease of the pylorus, so that food is retained and undergoes fermentation.

Yeasts (fig. 61, *e*).—v. Jaksch (*l.c.*, p. 108) describes three forms of yeasts as occurring in the vomit.

(*a*). Saccharomyces cerevisiæ. They are about the size of leucocytes and cohere in small groups. If a solution of iodine in iodide of potassium be run under the cover-glass, they stain a brownish-yellow. These bodies are occasionally elliptical in shape.

(*b*). Very small yeast-like fungi in thick clusters.

(*c*). Rod-like bodies, highly refracting, and of considerable length and thickness. They have rounded extremities, and occur separately or in strings. It has been stated that these are the agents in lactic acid fermentation of sugar.

Micro-organisms (fig. 61, *f*) are very frequently present in varying numbers. They take the form both of rods and cocci, but are non-pathogenic and therefore of no diagnostic worth.

Animal parasites.—These are purely accidental, and have entered the stomach from the intestines. The only ones which have been described are the Ascaris lumbricoides, Anchylostoma duodenale, and Oxyuris vermicularis.

CHAPTER XIX.

EXAMINATION OF DISCHARGES AND CONTENTS OF CAVITIES.

THE vast proportion of specimens which medical men are called upon to examine with the microscope, are derived either from the sputum or urine. But occasionally they have to investigate deposits of fluids drawn from the chest or abdomen, or the discharges from abscesses, fistulæ, &c. These materials, though hitherto commonly neglected, may yield valuable assistance in diagnosis, and the student therefore should make himself acquainted with at least the most important of such fluids, whenever occasion offers for their study.

The fluids will either have been drawn off by puncture, or in some other way, or a discharge may have occurred spontaneously. If sufficient quantity of the material can be obtained, it should be placed in a conical glass, and allowed to settle for some hours, when the supernatant liquid can be poured off, and portions of the sediment removed by means of a pipette for investigation. Unstained specimens should always be examined, and subsequently such staining processes, as experience may suggest, should be applied.

SEROUS EXUDATIONS.

These resemble in general appearance blood serum. They are almost clear, have a pale yellow colour, and coagulate on standing, yielding a clot which is rich in fibrin. As a rule there is but very little sediment, and when placed under the microscope in the usual way, it is often very difficult to make sure that one is focusing the layer of fluid between the glasses. A few red corpuscles will usually be seen, and they generally preserve their form and colour. Leucocytes are usually more numerous. Some endothelial cells derived from the surface of the serous membranes will be seen, but these are also scarce. In addition to these, Bizzozero (" Microscopie Clinique," p. 115) describes some larger cells, varying in diameter from 7 to 30 μ ; they contain small grey droplets, which are sometimes so abundant that the cells appear to be entirely made up of them. These bodies seem to undergo a kind of cystic degeneration, so that large vacuoles are formed, containing a clear fluid. Micro-organisms are frequently present, chiefly micrococci, and large numbers may be present although the fluid is quite clear.

In one specimen which I examined, the liquid was yellowish in colour and perfectly clear, with the exception of a few small white flakes. Yet tubercle bacilli were found in large numbers, especially in the flakes just mentioned. In order to search for the organisms, a little of the fluid is distributed over several coverglasses, which are then examined in precisely the same manner as films of sputum (see p. 281).

If the serous fluid is of old standing, cholesterin crystals are generally seen.

HÆMORRHAGIC EXUDATIONS.

Such liquids are of a brownish or red colour, due to the presence of blood. In the deposit red blood corpuscles, more or less altered in colour and appearance, are seen in large numbers, but occasionally the hæmo-globin seems to have been dissolved out, and then only a few colourless rings are visible. Endothelial cells, often with advanced fatty degeneration, are likewise to be found. In regard to these bodies, Quincke ("Deutsches Archiv für Klin. Med.," 1882, 569) asserts that when treated with a solution of iodine in iodide of potassium, a marked glycogen reaction is often produced, and that the probable presence of carcinoma is then indicated. Should they be accompanied by large epithelial cells, the diagnosis of cancer may safely be made.

CHYLOUS EXUDATIONS.

Beyond the appearance of particles of fatty matter no characteristic bodies can be found in the deposit of the fluid in such conditions.

SERO-PURULENT EXUDATIONS.

These often occur in tubercular cases, and the pathogenic organisms can be found without difficulty. Pus cells are of course numerous, but otherwise no special characteristics can be mentioned.

PUTRID EXUDATIONS.

These are usually brownish in colour, and are at once recognised by their penetrating and disagreeable odour. The deposit is peculiar, exhibiting in addition to shrunken red and white blood cells, numerous crystals, chiefly fatty needles, cholesterin and hæmatoidin. These fluids swarm with septic micro-organisms.

PURULENT EXUDATIONS.

If pus be allowed to stand in a cool place it separates into two layers, the upper of which is almost transparent, whilst the lower is whitish and opaque. A few streaks of blood are almost always present. When fresh, pus is of a greyish colour, but when putrid it appears like a thin green fluid with an extremely strong and unpleasant odour. If much blood be present the colour depends upon the proportion of the two fluids. The discharge from hepatic abscesses has a peculiar chocolate colour.

In the sediment of a purulent exudation pus cells of course almost fill the field ; if perfectly fresh, amœboid movements may be seen, and if a solution of iodine in iodide of potassium is run under the cover-glass, the corpuscles assume a mahogany colour. If the discharge is old, white cells, more or less shrunken and granular in appearance, are seen. Cells containing fat will also be noticed. Red blood corpuscles are usually present, sometimes in very large numbers.

Crystals are seldom absent, the most common being needles of the fatty acids, cholesterin, hæmatoidin and triple phosphates, but no special significance can be attached to their presence.

The search for **micro-organisms** is of the greatest importance; this is obvious, from the fact that the formation of pus has been shown to be almost entirely due to the action of bacteria for if these be excluded from entering the fluids of the body, as is now accomplished by antiseptic surgery, a purulent process never occurs. Certain organisms therefore are peculiar to pus ; these are various forms of micrococci. They are more numerous in old than in fresh pus, although they can also generally be detected in the latter. No fewer than eight varieties of micrococci have been separated from pus, but the most common are the following.

Staphylococcus pyogenes aureus.—These are cocci occurring singly, in pairs, in very short chains, and in irregular masses. Their presence causes no odour in the pus. They are found in the discharge from boils, and in the abscesses of pyæmia and puerperal fever.

Staphylococcus pyogenes albus.—Morphologically, these are indistinguishable from the above and occur under similar conditions, but differ somewhat in cultivations. Various pathogenic organisms are found in purulent discharges, and these are of the utmost importance, as their detection will be a most valuable guide to the surgeon or physician in the future treatment of the case.

Tubercle bacilli.—The method of detecting these organisms is the same as that described in the chapter on sputum. The layer of pus on the cover-glasses must be made as thin as possible, and to avoid errors it is important to allow the glasses to remain in the acid until the red fuchsine is entirely got rid of. Ten minutes or a quarter of an hour are not at all too long. The lesions in which the detection of tubercle bacilli is of great importance are many, amongst which may be mentioned tubercular nodules of the skin ; I have seen two or three cases in which so-called "post-mortem warts" were proved to contain bacilli.

Discharge from the ear in cases of tubercular meningitis in children is another example.

Actinomyces.—This is a rare disease, although cases seem to have been more numerous during the past few years, probably owing to our improved means

of investigation, and partly, perhaps, because a sys-
tematic examination of purulent discharges in doubtful
cases is more frequently made, than was formerly the
case. It should be laid down as a rule that when a
collection of fluid is diagnosed, if the nature of the disease is doubtful and especially if the discharge is purulent, a microscopic examination should always be made.

In actinomycosis the pus is thin and viscous, and
will be seen to contain small nodules of a greyish colour.
If one of these be placed under the microscope, the
characteristic club-shaped bodies will be seen.

Cover-glass preparations should be made in the usual
way and stained by the method described on p. 100.
The central mycelium will then be stained purple and
will be seen to consist of closely interwoven threads
surrounded by the radiating mass of clubs. This appearance is unique, and the "ray fungus" cannot be
mistaken for any other. In a case under Dr. Douglas
Powell at the Brompton Consumption Hospital Mr.
Taylor recognised the fungus in the discharge during life; and Prof. Crookshank undertook an exhaustive enquiry into the nature of the parasite, the
results being published in the Medico-Chirurgical Transactions, vol. lxxii., p. 193. He showed that the fungus
appeared to be closely related to the Basidio-mycetes,
and was able to demonstrate that the threads of the
mycelium pass directly into the narrow central ends of
the clubs.

Bacillus of leprosy.—The nodules which appear
on various parts of the skin and mucous membrane in
leprosy occasionally ulcerate, discharging a considerable quantity of thin pus. If cover-glass preparations
of this discharge are made in the usual way, and
stained in the same manner as sputum is stained for

tubercle bacilli, numerous fine slender rods will be seen, hardly distinguishable morphologically from the tubercle bacilli.   They stain, however, more readily, and unlike the bacilli of tubercle are not coloured by Kühne's method (see p. 92).

Bacillus of anthrax.—In this country two affections are produced by this organism.   " Malignant pustule," due to a direct inoculation, and consisting of a local inflammation followed by the development of a brawny swelling ; " wool-sorter's " disease resulting from inhalation, or swallowing of the spores so that the pulmonary or intestinal mucous membranes are first affected.

In order to demonstrate the pathogenic micro-organism, the local affection or carbuncle should be incised, and cover-glass preparations made of the discharge. When stained by Gram's method, the bacilli can be well seen as straight rods, sometimes curved.   Their extremities are sharply rectangular, and the bacilli are usually joined together in chains, the segments being marked off by clear linear spaces.   In sections the organisms should be stained by Weigert's modification of the Gram method as described on p. 96.

Bacillus of syphilis.—The pathogenic character of this organism has not met with general acceptance. It was first described by Lustgarten.   In form it somewhat resembles the tubercle bacillus, but it is always found in the interior of nucleated cells about twice the size of leucocytes.   They have been found in the discharge of the primary lesion and in hereditary affections of tertiary gummata.   Cover-glass preparations of the pus should be stained in the following way :—

A fresh solution of gentian aniline violet is made (see p. 95) in which the sections are placed for from

x

twelve to twenty-four hours. They are then removed, and washed for a few minutes in absolute alcohol, and afterwards immersed for ten minutes in a 1·5 per cent. solution of permanganate of potash. Next they are rinsed in a watery solution of pure sulphurous acid, and finally washed in water. If by these means the preparations are not completely decolourized, they must again be washed for a few seconds in permanganate of potash, and then again in sulphurous acid and water, the process being repeated until no more colour is visible. This method is not very satisfactory as it stains other bacilli, as well as those of syphilis, and the whole matter requires further research before it can be determined that the characteristic organ of syphilis has been found.

Another method has been introduced by De Giacomi which consists in staining cover-glass preparations in aniline fuchsine made in the same way as methyl aniline violet. The stain is heated, and the glasses allowed to remain in it for about 10 minutes, after which they are washed in water containing a few drops of perchloride of iron. They are next decolourized in a strong solution of perchloride of iron and counter-stained in Bismarck brown. The bacilli are then coloured a deep red.

Gonococcus.—The discharge obtained from patients suffering from gonorrhœa may be shown to contain a distinctive micro-organism in the form of micrococci, occurring either singly, or more usually in groups of ten or twelve. Cover-glass preparations of the pus are stained in methylene blue, and then thoroughly washed in water.

Bacillus of glanders.—The bacillus mallei is morphologically very similar to the tubercle bacillus.

It may be found in the pus derived from the ulcerated nasal passages of patients affected with this disease. The organisms are difficult to stain. The process introduced by Löffler is the best. The staining solution has the following composition :—

Concentrated alcoholic solution of
  methylene blue . . . . 1 part
Potash solution (1 in 10,000) . . . 1 part.

The potash solution must be added immediately before use.

Cover-glasses are immersed in the mixture for five minutes, and afterwards washed in water containing two or three drops of acetic acid. Next the preparations are transferred to another mixture containing two drops of concentrated sulphurous acid, and one of a 5 per cent. solution of oxalic acid in 10 c.c. of water. The alkaline staining fluid is thus decomposed, and the colour removed from everything except the bacilli.

Although not occurring in discharges, two other parasites may be mentioned here, both being more common in the flesh of cattle than in man, but, nevertheless, it is important to be able to recognise them.

Trichina spiralis (fig. 62).—This nematode has already been referred to on p. 246. In its characteristic form it is best seen in sections of muscle. It then appears coiled up in a lemon-shaped capsule, lying parallel to the muscular fibres, and surrounded at each end by a collection of fat globules. In some specimens the worm can hardly be distinguished, looking like a black mass, and the capsule will then be found to be calcified.

Cysticercus cellulosæ.—This has in rare cases been found in man. A case is recorded by Dr. Betham Robinson in the "Transactions of the Pathological

Society," 1889, p. 388, as occurring in the trapezius muscle of a child, aged four. Most cases that have been reported have been in the eye and brain, where its presence is most conspicuous, but in animals it is found chiefly in the muscles. It occurs as a small round body

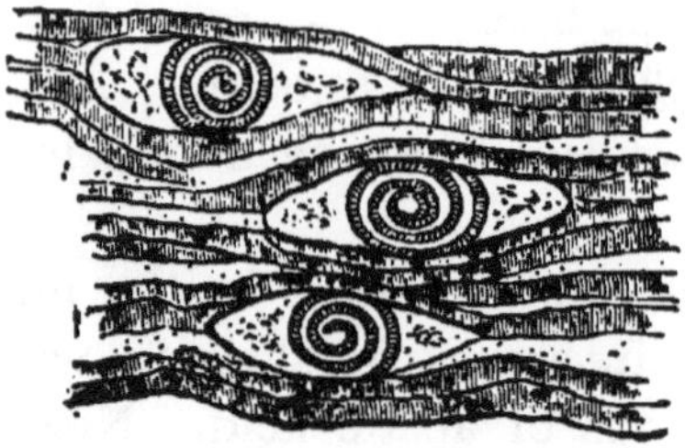

FIG. 62.—Trichina spiralis in muscle (Parkes).

visible to the naked eye, being sometimes as large as ¼ inch in diameter. Under the microscope the head (fig. 63) will be seen to have four suckers and a circle of hooklets.

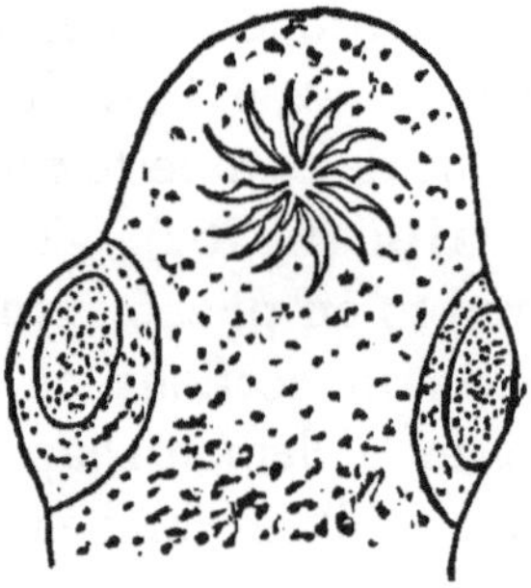

FIG. 63.—Head of cysticercus cellulosæ (Parkes).

CONTENTS OF CYSTS.

For the purpose of diagnosis the contents of cysts are usually obtained by means of an exploring needle, but if possible it is better to procure a larger quantity by aspiration. The fluid is allowed to stand for some hours, and then some of the deposit is removed by a pipette, and examined by the microscope in the usual way.

Hydatid cysts.—The fluid obtained from these cysts is clear and limpid, with a specific gravity of about 1008. It contains very little albumen, and abundance of inorganic salts.

The deposit is characteristic; it contains the immature form of the Tænia echinococcus, a parasite

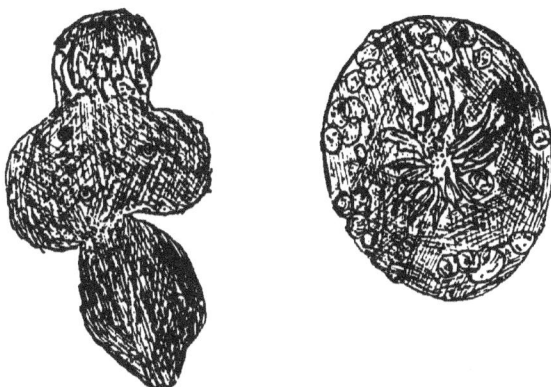

FIG. 64.—Hydatid cysts.

which inhabits the intestine of the dog. The ova are discharged with the fæces of the host, and in some way, probably through drinking water, find there way into the human stomach, and thence into some other organ, generally the liver. There they develop and form minute round bodies known as scolices (fig. 64). In the sediment these can hardly be seen with the naked eye. They are provided with two circles of hooklets and four suckers. Each scolex is oval in form, transparent, and often studded with granules, which are frequently stained yellow by the bile. In addition, free hooklets will also be seen scattered through the specimen. In shape these resemble tiger's claws.

In most specimens of this parasite, the original cyst formed from the ovum usually developes another within itself, known as the "daughter-cyst," and these again

FIG. 65.—Wall of hydatid cyst (Boyce).

develop still more, "granddaughter-cysts." In this way the echinococcus becomes filled with a number of small vesicles of various sizes, which may amount to thousands. This illustrates the difference between a "cysticercus" and an "echinococcus," whereas the former gives rise to only one scolex, and can ultimately form only a single tapeworm, the latter may develop thousands of both.

Occasionally the parasites cannot be found, but only portions of the cyst wall (fig. 65) or "cuticula." This is made up of a number of concentric layers, composed of a chitinous material which has a very characteristic appearance, and when once seen cannot be mistaken.

These cysts occasionally suppurate, and the hooklets then are generally the only recognisable feature that remain.

Ovarian cysts.—The fluid contents of ovarian cysts vary considerably in character. The fluid obtained may be either perfectly limpid or of a gelatinous consistence. The deposit as a rule is copious. The cells which are formed are but seldom well preserved, fatty degeneration being so far advanced that they are scarcely recognisable ; free nuclei are abundant. Epithelial cells may be either squamous, columnar, or ciliated. Colloid concretions which are probably derived from epithelium are an invariable component of "colloid cysts." Red and white blood corpuscles occur in varying numbers, and cholesterin crystals are also generally present.

Dermoid cysts.—The fluid in these cysts is thick and brownish in colour. The deposit is very copious, containing, in addition to epithelium, hairs and various forms of crystals, especially fat, cholesterin and hæmatoidin.

Hydronephrosis.—The diagnosis of cystic kidney can be best assured by chemical tests, namely, those for the detection of uric acid and urea.

The deposit, though usually slight, is of importance, and must be carefully collected in a conical glass. On microscopic examination epithelium from the renal tubules must be sought for, and if found, a diagnosis of cystic kidney may safely be made.

Spermatocele.—The nature of such cysts is at once determined by the presence of spermatozoa (fig. 33, p. 184) and occasionally crystals closely resembling the Charcot-Leyden crystals.

CHAPTER XX.

DISCHARGES FROM THE GENITAL ORGANS.

A MICROSCOPICAL examination of discharges from the genital organs often yields important information, especially from a medico-legal point of view. A medical man may be called upon to decide as to whether certain stains on linen are spermatic or not, or whether a piece of membrane is derived from the decidua of an early abortion. In this chapter a brief description will be given of the most important objects seen under the microscope in the examination of such discharges. . For full details and the conclusions to be derived from them, the reader is referred to treatises on Medical Jurisprudence and Obstetrics, although I have been unable to find any accurate descriptions of the various forms of membrane which may be discharged from the female genital organs.

DISCHARGES FROM THE MALE GENITAL ORGANS.

Seminal fluid.—Semen, when fresh, is of a viscid, honey-like consistence, somewhat opaque, of slightly alkaline reaction, and possesses a faint characteristic odour. As it cools it becomes temporarily gelatinous, and afterwards changes into a thick liquid. Its normal colour is white, but under various conditions this may be altered.

If mixed with blood it may appear red, brown, or

yellow. The blood may be derived from the prostatic urethra, in which case spots of pure blood are usually noticed in addition to the brownish colour. If the hæmorrhage come from the vesiculæ seminales, the spots are all equally coloured, denoting an intimate mixture of blood and semen (Ultzmann, " Sterility and Impotence," p. 15).

A yellow colour is usually due to the admixture of pus, and the stains on linen are then greenish or yellowish.

Occasionally indigo is present, and a violet or claret tinge is produced.

Under the microscope the most prominent objects are the **spermatozoa** (fig. 66, *a*). In normal conditions these are present in large numbers. In fresh semen they are seen in active movement. They are about fifty μ in length. The "head" is pear-shaped, and 4·5 μ long. The tail is tapering, if a little water is added it will be seen to roll up and form a loop. If the organisms are dead, the tail will be seen to be spirally coiled or bent at an angle. Their characteristics are very well displayed by the addition of a drop of iodine solution.

A question of sterility may be raised, when the microscope may be of great value.

Distinction must be drawn between Aspermatism and Azoospermism.

By **aspermatism** is meant a condition in which there is inability to ejaculate semen. It may be permanent or temporary, congenital or acquired.

In **azoospermism,** although fluid may be ejected, no spermatozoa can be found, and the individual in question therefore is sterile. This condition also may be permanent or temporary, congenital or acquired.

The spermatozoa may be greatly diminished in

number without being quite absent, this is known as **Olizoospermism.** Such a condition is common in advanced age, or may be produced by disease, such as gonorrhœa.

As long as the spermatozoa can be seen to exhibit active movements, the patient cannot be said to be absolutely sterile, although impregnation is improbable.

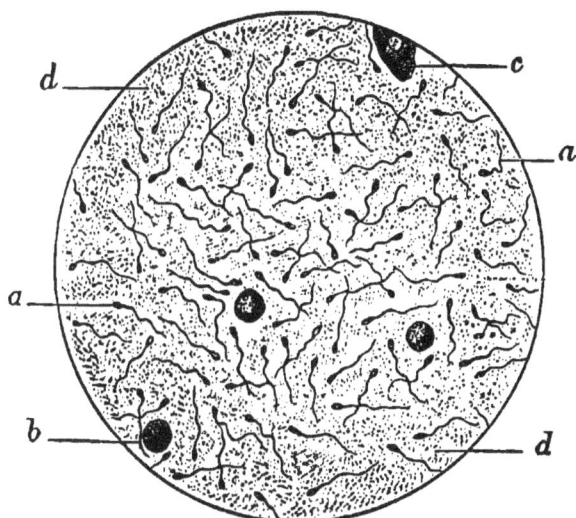

FIG. 66.—Normal semen (Ultzmann). *a.* Spermatozoa. *b.* Seminal cells. *c.* Epithelium. *d.* Seminal granules.

In addition to spermatozoa, certain other bodies are seen under the microscope in semen.

Seminal cells (fig. 66, *b*).—These are circular in shape, finely granular, and represent the spermatozoa in an embryonic state; they usually exhibit one or more nuclei.

Epithelium (fig. 66, *c*).—These are of the squamous or columnar varieties.

Masses of **amyloid substance,** which are finely

granular within, stratified, and enclose a central nucleus. These are derived from the prostatic secretion (von Jaksch).

Seminal granules (fig. 66, *d*).—These form a fine molecular débris. They are often found attached to the seminal casts, which have been already described on page 183.

Spermatic crystals.—In general form these bodies closely resemble the Charcot-Leyden crystals (see page 275). Many authors consider them to be chemically the same. Fürbinger asserts that they are derived entirely from the prostatic secretion. In normal semen they are not deposited for several hours and are seen to be most numerous and most perfectly formed in cases of azoospermism.

Spermatic stains.—When semen has dried on linen or cotton, the fibre of the stuff is stiffened, and the stain has a slightly translucent appearance.

In order to demonstrate the nature of the stain small pieces of the linen should be carefully separated and placed in a watch glass with a few drops of water, just sufficient to soak the fibre thoroughly.

The watch glass should be covered to keep out dust and prevent evaporation, and set aside for an hour. Afterwards some of the fibres are removed and gently pressed on clean glass slides. A slightly turbid fluid will exude, which can be rendered clearer, without injury to the specimens, by the addition of a little dilute acetic acid. On application of cover-glasses, and examining the preparations under a quarter inch lens, the dead spermatozoa will be found if the stain was caused by semen. Observation should at the same time be made of accompanying hairs or fibres of dress stuff, &c. If too much fluid has been added the spermatozoa may become so transparent as to be almost invisible. One or two specimens should, therefore, be allowed to dry by evaporation, and a couple of drops of iodine solution added to the sediment, the organisms will then be well seen.

Discharges from the genital organs in cases of **gonorrhœa** and **syphilis** may be examined for the characteristic organisms.   The methods for so doing have been described on pages 305 and 306.

DISCHARGES FROM THE FEMALE GENITAL ORGANS.

The **vaginal secretion** under normal circum-stances exhibits under the microscope squamous epi-thelium cells and a few large leucocytes.

When the leucocytes are very numerous, vaginal catarrh is probably present, and a few red blood cells will be seen mixed with the white corpuscles.

Cancer cells.—In cases of cancer affecting the vaginal portion of the os uteri, in which there is ulcera-tion, large epithelial cells, irregular in shape, and con-taining large nuclei, are stated to be found in large numbers, and to be true "cancer cells," but it is very doubtful, and whether they can be distinguished from the epithelial cells of the vagina, &c., I am much in-clined to agree with those who consider "cancer cells" to be a myth.

Parasites:—

Trichomonas vaginalis.—In chronic catarrh of the vagina, in addition to a large quantity of epithelial cells, and mucous and pus corpuscles, an infusorium is found, known as the "trichomonas vaginalis" (fig. 67). It is oval in shape, and has a long caudal appendage, three flagella and a lateral row of cilia. In the specimen from which the accompanying sketch was taken, the cilia could not be seen.

Fungi.—Numerous micrococci are nearly always found in the vaginal secretion, they are greatly increased after childbirth.

Tubercle bacilli have occasionally been found; cover-glasses should be prepared and stained in the usual way (see p. 28).

Gonococci (see p. 306) are found in patients suffering from gonorrhœa.

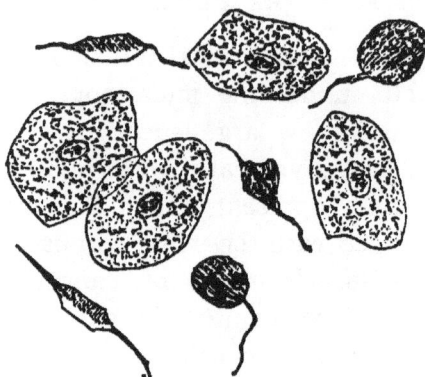

Fig. 67.—Trichomonas vaginalis.

In cases of suspected rape, the vaginal mucus should be examined for spermatozoa.

Membranes.—Various forms of membrane may be passed from the female genital organs, and the microscope may be of great aid in helping to decide the nature of the case.

Mucus coagulated by astringents may simulate a membrane. By floating it out in water its true composition will probably be recognised; if placed under the microscope, a large number of cells about the size of leucocytes will be seen variously altered in shape, and irregularly distributed amongst ill-defined fibres. When an attempt is made to tease the material out, it breaks up into fragments, so that it will be found almost impossible to obtain a thin layer, and the component parts do not adhere together as is the case with true membrane.

Fibrinous clots.—These may closely resemble the menstrual decidua of membranous dysmenorrhœa. On floating in water, however, the clot is easily broken up. Under the microscope red blood cells are seen in abundance together with some fibres, the former constituting the chief bulk of the specimen.

Vaginal casts.—These are easily recognised with the microscope. They consist almost entirely of squamous epithelial cells closely packed together, amongst which a few leucocytes and red blood cells may be seen. Their characteristics are best demonstrated by the addition of a couple of drops of a two per cent. solution of methylene blue.

Membranous dysmenorrhœa.—In this condition, during the menstrual period, generally on the second or third day, membrane is expelled either in shreds or forming a more or less complete cast of the uterus.

Under the microscope, layers of fibrin enclosing red blood corpuscles are seen; one surface is rough and uneven, whilst on the other small round cells embedded in a reticular net-work may be made out, amongst which a few tubular channels are noticed; these are the apertures of uterine glands, and are therefore of high diagnostic value. The epithelium lining the ducts is rarely complete but is usually more or less destroyed.

In complete casts the orifices of the Fallopian tubes

can be demonstrated, and the whole structure of the mucous membrane is seen including enlarged orifices of glands and an unusual amount of fibrillar tissue.

Membranes of early abortion.—The recognition of an early abortion is often attended with difficulty; when the decidua have fully formed it is of course easier, but if the event occurs within the first few months a microscopic examination may often be of great value. Portions of the amnion only may be found. It is a tough membrane, perfectly smooth and transparent. The internal surface is smooth and glistening and consists of a layer of flattened cells, each containing a large nucleus. This layer rests on a stratum of fibrous tissue, which gives the membrane its firmness, and is attached to the inner surface of the chorion. No vessels, nerves or lymphatics can be seen in it. As a rule, the amnion and chorion are found together. One surface of the membrane is then smooth— the amnion—and the other rough owing to the presence of the chorionic villi.

If the membrane be teased out and examined under the microscope the following structures can be distinguished, although their exact distribution will probably have been disturbed.

1. A large number of small round cells.

2. A certain amount of fibrous tissue.

3. Irregular spaces filled with red blood corpuscles, many of them looking like large blood vessels.

4. Flattened cells with large nuclei, these belong to the amnion.

5. Whorl-like collections of young connective tissue cells. These formations are circumscribed, and consist of delicate branching fibres with nuclei. They are the chorionic villi. Their appearance is very characteristic

and when once seen can hardly be mistaken. I have not found anything precisely similar to them in any other structure, and I think one would be safe in diagnosing fœtal membranes when they are present.

6. A few cubical cells with large nuclei, the so-called " lentil-shaped " cells (Virchow). These are derived from the decidua.

If the decidua are also seen the arrangement of the layers is usually more distinct and have the following arrangement :—

On one surface are the flattened cells of the amnion and beneath is the young connective tissue of the chorion.

Next comes a thick layer of the "lentil-shaped" cells, those nearest the fœtal membranes being closer together and smaller than those nearer the uterine surface. Then a certain amount of glandular tissue will probably be seen, the epithelium lining the ducts being often perfectly intact and finally some muscular bundles detached from the uterus ; the characteristic portions being the " lentil-shaped " cells just alluded to.

CHAPTER XXI.

Examination of the Blood.

A microscopical examination of the blood is usually undertaken for the purpose of ascertaining the relative proportion in number of the red and white corpuscles to the normal, and to one another. In addition, certain changes in form of the red corpuscles are of importance and the amount of colouring matter is also usually estimated. More rarely, various micro-organisms are sought for.

In normal blood the proportion of white to red cells is about 1 to 300, this varying considerably within physiological limits; thus, the white cells are greatly increased in number after a meal when the proportion may rise to 1 to 150, afterwards falling to 1 to 600.

The **red corpuscles** are $\frac{1}{3200}$ inch ($7\cdot5\ \mu$) in diameter, and $\frac{1}{12000}$ inch ($2\ \mu$) in thickness. When seen sideways they are bi-concave or dumb-bell-shaped and of a pale buff colour which appears of a reddish tint when a number of corpuscles are collected together. They have a tendency to run together, forming rolls or rouleaux. They are non-nucleated.

When the circumference of a corpuscle is brought sharply into focus, a darker area is seen in its centre, but if the lens is brought nearer the corpuscle by means of the fine adjustment, the darker centre is replaced by a lighter area, while the ring becomes darker; this being due to the fact that these bodies are bi-concave, circular discs.

The **white corpuscles** are usually spherical in form and faintly granular. If the blood is examined fresh on a warm stage, amœboid movements may be noticed. Their average size is about $\frac{1}{2500}$ inch, or 10 μ in diameter. One or more nuclei are present, but are usually not visible unless a little weak acetic acid is run under the cover-glass.

A third element will sometimes be observed, namely, **blood-plates.** These were first noticed by Bizzozero. To demonstrate them the blood must be " fixed " by the addition of some preservative fluid, such as Hayem's. This has the following composition :—

Chloride of sodium.	1 part
Sulphate of sodium	5 parts
Corrosive sublimate	·5 parts
Distilled water	200 parts.

A couple of drops of this fluid is added to a drop of the blood obtained from the finger, and placed on a glass slide. A cover-glass is applied, and the specimen examined by the highest obtainable power (preferably a $\frac{1}{12}$ oil immersion) and a small diaphragm. The plates will then be seen as slightly oval or rounded granules about one-third the size of the ordinary red corpuscles.

Methods of examination.—For the histological examination of the blood it is sufficient to tie a handkerchief tightly around one finger so as to compress as much blood as possible into the tip, then prick it with a sharp needle, and remove the drop which exudes on a clean cover-glass. This should be placed on a glass slide and its edges surrounded with a rim of oil or Canada balsam, in order to prevent evaporation.

Two processes of the utmost importance in the examination of the blood will now be described, *i.e.*, " counting the corpuscles," and the " estimation of hæmoglobin." The latter does not strictly come under the head of Medical Microscopy, but a chapter on ex- amination of the blood for clinical purposes can hardly be complete without it.

COUNTING THE CORPUSCLES.

This is usually accomplished by means of Dr. Gowers' **hæmocytometer,** fig. 68, (made by Hawksley, Oxford Street), which consists of (1) a pipette graduated to 995 cubic millimetres for measuring the diluting solution ; (2) a capillary tube for measuring the blood and capable of holding 5 cubic millimetres ; (3) a small glass jar and and stirrer for making the dilution ; and (4) a glass slide on which is arranged a cell ·2 millimetres deep, and ruled at the bottom in squares, each ·1 millimetre in length and breadth. The slide is attached to a metal plate furnished with two springs for maintaining the cover-glass in position.

The diluting solution recommended by Dr. Gowers is as follows :—

Sulphate of soda . . . . . 104 grains
Acetic acid . . . . . . 1 drachm
Distilled water . . . . . 6 ounces.

The instrument is used in the following manner :—- 995 cubic millimetres of the saline solution are mea- sured off in the pipette and placed in the mixing glass. A drop of blood is obtained from the patient's finger in the way described above, and drawn up by the capillary tube to the level of the 5 cubic millimetre mark, and then thoroughly mixed with the saline solution in the glass. One caution must here be given, and that is that if the drop of blood does not readily exude, too much pressure must not be applied, or otherwise too great a proportion of serum will be obtained, and thus an incorrect calculation will result.

The diluted blood is then thoroughly stirred with the small glass rod; one drop is transferred to the floor of

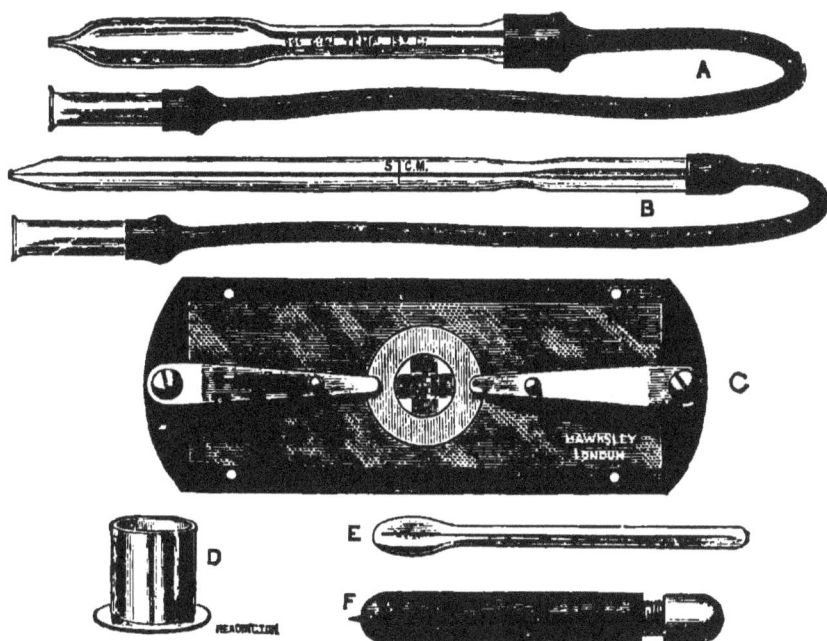

FIG. 68.—Gowers' hæmocytometer. A. Pipette for measuring the diluting solution. B. Capillary tube for measuring the blood. C. Cell with divisions on the floor. D. Vessel in which the dilution is made. E. Spud for mixing the blood and solution. F. Guarded spear-pointed needle.

the cell, and another to a cover-glass. The two drops are then approximated, which ensures a regular layer extending throughout the depths of the cell, and the springs are arranged so as to keep the cover-glass in position. It is a good plan before placing the blood in the cell to rub the ruled portion of the cell over with a soft black-lead pencil, afterwards removing the loose dust. This renders the squares distinct, otherwise, especially in a new glass, they are often barely visible (Wynter).

The slide is now examined with a quarter-inch lens, and the number of corpuscles counted in 10 squares. In order to explain the computation we quote Dr. Gowers' own words in his article in Quain's "Dictionary of Medicine."

"The dilution of 5 cmm. of blood in 995 cmm. of solution is 1 in 200; each square contains the corpuscles from a volume of dilution ·2 mm. in one, and ·1 mm. in each of the other dimensions—that is, 2 cubic ·1 mm., or the ·002 part of a cubic mm. But the dilution being 1 in 200, this volume of dilution contains just ·00001 cm. of blood. The number of corpuscles in a square multiplied by 100,000 is thus the number in a cubic millimetre of blood—the common mode of statement. In order to limit error, the total number of corpuscles in 10 squares should be counted, and this number multiplied by 10,000 is the number per cubic millimetre."

The average number in normal blood is about 5,000,000, which corresponds to about 50 in a square; from this it is evident that normal blood will contain 100 corpuscles in 2 squares, and that, therefore, the number of corpuscles of any blood in 2 squares of the hæmacytometer represents the percentage of corpuscles compared with normal blood.

In counting the white corpuscles, if they are not in considerable excess, it is most convenient first to ascertain the number of red corpuscles per square, and note how many squares are contained in the field of the microscope. If then the focus is raised so that the corpuscles gradually become indistinct, the white ones, from their higher refracting power, will appear like bright points, and the number in a series of fields can easily be counted. For example, the number of red corpuscles per square has been found to be 40, and the field contains 15 squares, that is, 600 corpuscles per field. 10 fields contain 15 white corpuscles ; the proportion of white to red will, therefore,

be 1 to $\dfrac{600 \times 10}{15} = 1$ to 400.

Another excellent instrument, though not so well-known in this country is the *Thoma-Zeiss hæmocytometer* (may be procured from Baker, 243 High Holborn). Dr. Archibald Garrod has had considerable experience in the use of this apparatus, and I have to offer him my best thanks for kindly writing for me a description of the hæmocytometer, and the mode of using it.

The dilution and mixing of the blood with saline solution is carried out in a pipette (fig. 69) made after the pattern of Potain's melangeur. The capillary bore of this pipette is at one point expanded into a bulb in which lies a small glass ball. The lower portion of the tube is graduated, and into it are drawn 5 or 10 cubic millimetres of blood ; saline solution (of the composition

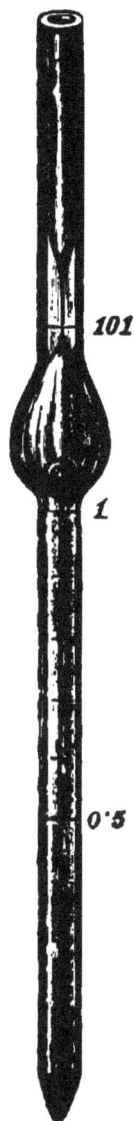

FIG. 69.—Pipette (Thoma-Zeiss hæmocytometer).

given above) is then drawn in until the bulb, and a very small portion of the capillary tube beyond, are filled with the mixed blood and solution. According to the amount of blood taken, the dilution will be 200 or 100 times. A gentle rotary motion is then imparted to the pipette, which, by causing the glass ball to move within the bulb

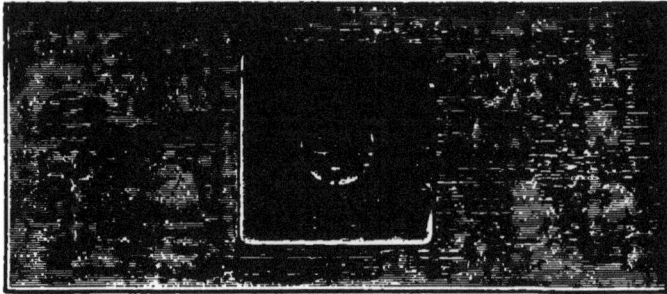

FIG. 70.—Counting slide (Thoma-Zeiss apparatus).

brings about a complete mixture of its contents. A small portion of the liquid is then expelled so as to clear out the unmixed saline solution from the capillary tube, and afterwards a drop of the diluted blood is placed upon the counting slide (fig. 70). The centre of this slide is formed into a circular cell by a rim of glass, and

FIG. 71.—Cross section of Thoma-Zeiss apparatus.

in the middle of the cell so formed is a small platform not quite so thick as the surrounding rim (fig. 71). When the cover-glass is placed over the cell, a space of ·01 millimetre is left between it and the central platform, whilst the small trough between the platform and the rim serves as a receptacle for any overflow of fluid.

When the cover-glass is accurately applied, Newtonian colours will be produced where it is in contact with the rim of the cell. When these colours cannot be obtained, it must be concluded that some dust is lying between the two surfaces of glass which must then be separated and cleansed, and a fresh drop of liquid from the pipette placed in the cell. The preparation must also be re-jected if any of the liquid makes its way between the cover-glass and the external rim.

Upon the centre of the platform is a square divided

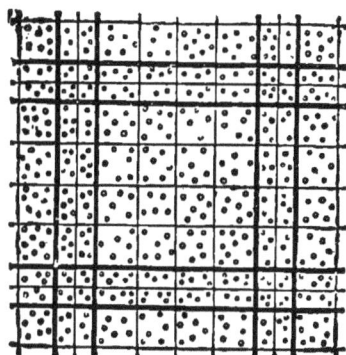

FIG. 72.—Counting scale (Thoma-Zeiss apparatus).

into 400 lesser squares, each $\frac{1}{400}$ square millimetre in area and $\frac{1}{20}$ mm. in diameter. Since the depth of cell is ·01 c.mm. the space over each square is $\frac{1}{4000}$ c.mm. The contents of 16 such squares gives the per-centage of corpuscles. Sixteen sets of sixteen squares are marked off by triple lines (fig. 72), and if the whole of these are counted, or 256 squares in all, the area included in the counting equals that of 32 squares of Gowers' hæmocytometer.

In order to estimate the number of red corpuscles per cubic millimetre of undiluted blood, if after counting the

corpuscles in a certain number of squares, m be found to
be the average number of corpuscles per square of $\frac{1}{400}$
millimetre area, this number would have to be multiplied
by 4000 × .100 or 200 (according to the dilution taken)
to give the result sought for.   Or the calculation may
be performed thus :— $\dfrac{4000 \times 200 \text{ (or 100)} \times x}{y} = n,$
where $x =$ total number of corpuscles counted, $y =$ the
number of squares and the result (n) will be the number
of corpuscles in a cubic millimetre of undiluted blood.

Dr. Garrod says: "I always count the squares in sets
of 16, counting successive rows of four squares, including in my counting those corpuscles which touch the
upper and left hand borders of the square, either from
the outer or inner side, and excluding those which touch
the lower and right hand borders."

The small size of the squares of this instrument very
greatly facilitates the process of counting ; which may
also be interrupted if necessary, and continued later on
from the point at which it was stopped, as a note can be
made of the number, and row of the square last counted.

One of the drawbacks to the instrument is the trouble
required for the cleaning of the pipette which should be
washed out several times with water, and then dried.
The drying is effected by means of alcohol, and after
the expulsion of the spirit a current of dry air should be
sent through the pipette until the little glass ball moves
freely about in the bulb, without adhering to the sides.
The pipette should also be washed out occasionally with
ether, and at intervals with caustic potash.

For the counting of the white corpuscles a second
pipette is provided, of similar form, but giving a dilution
of 10 or 20 times.   The liquid employed is a $\frac{1}{3}$ per cent.
solution glacial acetic acid, which destroys the red

corpuscles and at the same time renders the white more clearly visible. A drop of the mixed blood and acid is placed in the cell and the corpuscles are counted by fields of the microscope, the area of which is determined in the following manner.

The diameter of the field is ascertained by means of divisions on the floor of the cell, each of which equals $\frac{1}{20}$ millimetre. If the field has a chamber of ten divisions, the radius will be $\frac{1}{4}$ millimetre, and the area π $(\frac{1}{4})^2$ square millimetre, and since the depth of the cell is ·01 millimetre, the cubic contents of the field will be $0·1 \times \pi$ $(\frac{1}{4})^2$ cubic millimetre $(\pi = 3·1416)$. The number of white corpuscles may be then calculated by means of the following formula :—

$$\frac{D \times N}{F \times C}$$

when   D = the dilution *i.e.*, 10 or 20

N = number of corpuscles counted

F = number of fields counted

C = cubic contents of a field.

In practice the calculation may be greatly simplified by always counting the same number of fields (*e.g.* 50), and by always using the same sized field, and the same dilution of the blood.

ESTIMATION OF THE AMOUNT OF HÆMOGLOBIN IN THE BLOOD.

Various modes of estimating the richness of the blood in hæmoglobin have been introduced. It may be accomplished by the amount of dilution necessary to obscure a certain absorption band in the spectrum, or by the coloured discs invented by Hayem. The two

most practical methods for clinical purposes are those of
Dr. Gowers and von Fleischl.

Gowers' hæmoglobinometer (fig 73).—This apparatus consists of two small tubes of exactly equal
diameter, which when in use are placed in a stand made
to receive them.   One tube contains a preparation of

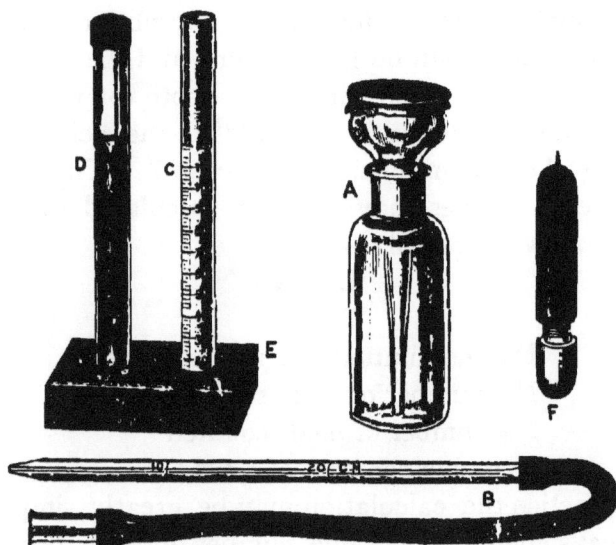

FIG. 73.—Gowers' hæmoglobinometer. A. Bottle for holding the
diluting solution. B. Capillary pipette for measuring the blood.
C. Graduated tube for measuring the amount of hæmoglobin. D.
Standard tint of normal blood. E. Support for D and C. F. Puncturing needle (Hawksley).

carmine and glycerine jelly coloured to represent a solution of 1 part of healthy blood in 100 of water.

The other tube is graduated, each division being
equal to the volume of blood taken (20 cubic millimetres).   So that 100 divisions equal 100 times the
volume of blood. A capillary pipette is also supplied
capable of holding 20 cubic millimetres.   This is filled

with blood which is blown into the graduated tube in which a few drops of distilled water have previously been placed. The blood and the water are then thoroughly mixed. Distilled water is now added drop by drop from the pipette until the tint of the mixture corresponds with that of the standard.

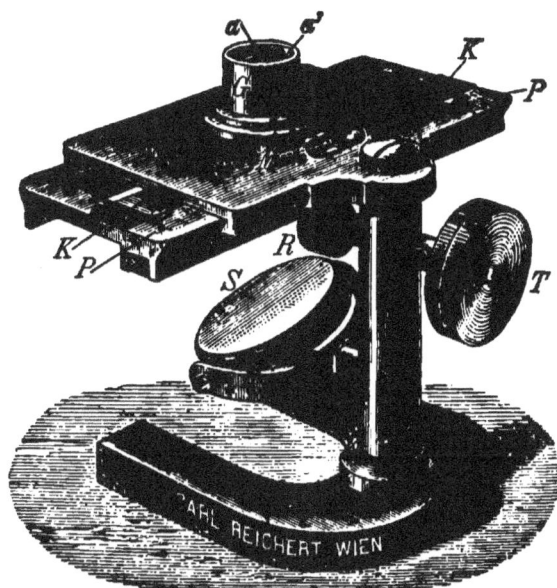

FIG. 74.—Von Fleischl's hæmometer (for explanation of lettering see text).

The height of the diluted fluid is then read off, the division which is reached representing the percentage amount of hæmoglobin.

For example, a specimen of blood is diluted until the mixed fluids reach the 55 mark on the tube, when the tint of the standard is found to be attained. The blood examined, therefore, contains 55 per cent. of the normal quantity of hæmoglobin.

Von Fleischl's hæmometer (fig. 74).—For an

account of this excellent piece of apparatus, I am once more indebted to Dr. Archibald Garrod, who describes it and the means of using it in the following words :—

" In this instrument the standard of comparison is an elongated wedge of tinted glass (K), which is so mounted that it can be moved backwards and forwards beneath a small stage, by turning a milled wheel (T). In the stage is a circular opening through which the light of a lamp is reflected upwards from a disc of plaster of Paris (S) mounted like the mirror of a microscope. The wedge of glass underlies one-half of this opening. The mixture of blood and water to be examined is placed in one division of a cylindrical vessel (G), which has a glass bottom, and is divided in half by a vertical septum (a). This vessel fits accurately over the circular opening in the stage, and must be so adjusted that the septum coincides with the edge of the tinted prism. On looking down upon the instrument, one-half of the illuminated circle will then receive its tint from the prism, the other half from the mixture contained in the vessel.

" The blood is received from the ear or finger into a minute glass tube, into which it is drawn by capillary attraction. This tube is held by a small metal handle, which afterwards serves as a convenient stirring rod. The tube being accurately filled, and any blood which may have adhered to its outer surface having been carefully removed, its contents are *immediately* washed out into the mixing vessel by means of a small pipette filled with water. Water is then added with careful stirring, until a plane surface is obtained, level with the top of the cylindrical vessel, and showing neither a concave nor a convex meniscus. If any clotting has occurred in the tube, a fresh specimen should be taken, but after a little practice this accident will seldom occur.

" The other half of the vessel, which lies over the tinted prism, must then be filled in the same manner with pure water. The comparison of the tints is facilitated by placing a cardboard tube, blackened internally, upon the stage of the instrument, so as to enclose the mixing vessel.

" The prism is then swept backwards and forwards until an accurate match is obtained, when the percentage of hæmoglobin may be read off upon a scale (P) which is visible through a second small opening (M) in the stage. It is well to repeat the comparison several times, until the successive readings exhibit a close agreement.

" The estimate should be arrived at as quickly as possible, because the eye soon becomes tired, rendering an accurate comparison almost impossible. In case of doubt the eye should be rested, and a fresh determination made after an interval.

" The comparison requires a darkened room and an artificial light, since by daylight the tint of the prism is entirely unlike that of the diluted blood.

" The standard of hæmoglobin is somewhat higher than that of Gowers' hæmoglobinometer, and I have not yet found the blood of any Londoner to attain to the 100 per cent.

" I have made repeated observations in quick succession, taking a fresh specimen of the blood of the same individuals, and have obtained results not differing by more than 1 per cent., but it should be mentioned that the scale is only divided into tens and fives, although I believe that the delicacy of the instrument would admit of finer graduation. As it is the units of percentage must be estimated by the observer."

Having described the methods of estimating the

number of the corpuscles, we must now proceed to consider what results may be arrived at from this computation. The red corpuscles may be decreased, this condition being known as **oligocythæmia.** As before stated the normal number of red corpuscles in the blood is 5,000,000 in a man, and 4,500,000 in a woman to the cubic millimetre of blood. In various forms of disease this number may be diminished, the number falling to 2,000,000 or even as low as 360,000 per cubic millimetre (v. Jaksch). The lowest number I have ever found was 450,000 ; this occurred in a case of pernicious anæmia.

Temporarily this decrease may be due to hæmorrhage. As a permanent condition it is found in those diseases which are attended with deficient production of the corpuscles.

The quantity of hæmoglobin is roughly proportionate to the number of corpuscles, but the individual corpuscles often contain less hæmoglobin than normal. The loss may be so great that the proportion of hæmoglobin falls to 25 per cent. of the normal, as in some cases of chlorosis. Hayem drew attention to the fact that in pernicious anæmia the number of red cells in the blood is inversely proportional to the quantity of hæmoglobin which they contain. In the case just referred to, the blood, tested by v. Fleischl's hæmometer only yielded 10 per cent. of hæmoglobin.

Polycythæmia.—Increase in number of the red corpuscles is never great, being very transitory, and within physiological limits, thus it is found after meals and in the newly born. In plethoric conditions also the number is above the average, and in the algid stage of cholera the red corpuscles are relatively in excess.

Leucocythæmia.—By this is meant a condition in which there is a considerable and permanent increase in the number of the white red corpuscles ; it is also known as *leucæmia*.

Microscopical examination sometimes shows an enormous increase in the number of the white corpuscles. Virchow has found them as high as in the proportion of 2 : 3 of red corpuscles. In this condition the red corpuscles are also diminished, being generally about two or three millions to the cubic millimetre of blood.

Von Jaksch (*l.c.*, page 20) has made some remarkable observations as regards the character of the leucocytes. It is usual to distinguish leucæmia as splenic, lymphatic, and myelogenic, according to the anatomical seat and clinical symptoms of the disease. When large and small leucocytes, the latter preponderating, occur, the leucæmia is of a lymphatico-splenic character. When the larger cells alone are found, v. Jaksch considers that the disease is almost entirely splenic. If the corpuscles are of a transitional form, nucleated red cells being also present, the bone-marrow is probably the seat of serious changes.

Although the more pronounced forms of leucocythæmia are recognised without difficulty, it is not so easy to detect an early stage of this condition.

Prof. Ehrlich has pointed out how this may be accomplished, by taking advantage of the selective properties of the " granules " present in the protoplasm of the leucocytes towards eosin.

Preparations are made as follows :—Two films of blood are procured by pressing a drop of blood between two cover-glasses, and then sliding them apart. They are next dried in an exsiccator and heated upon copper foil for a considerable time (ten to twelve hours) at 12c° to 130° C. A more simple expedient may be adopted by placing the glasses as soon as the blood has partially

z

dried in absolute alcohol for a couple of hours, and then driving off the spirit by heat. A drop of concentrated eosin-glycerine solution is then added to each cover-glass, the excess of stain is washed off with water, and the glasses are finally dried and mounted in balsam.

Ehrlich found that the granules presented considerable differences in their staining properties. He distinguished five several varieties, classifying them as α to ε granules.

When the disease is acute, and the number of leucocytes only slightly diminished, the mono- and polynuclear forms (ε-granules) are alone increased, whilst the α-granules (" eosinophile ") are decreased. When the condition is a more chronic one, exactly the reverse condition of affairs results, the α-granules being increased and the ε-granules diminished.

Various changes in the red corpuscles are noted in leucocythæmia, but these will be described separately.

Leucocytosis.—This term is applied to the condition in which the white corpuscles are temporarily increased. This occurs physiologically after a meal, the proportion of white rising as high as 1 to 100 red corpuscles. According to Virchow, leucocytosis accompanies every case of lymphatic excitement, such as inflammation, and tubercular, scrofulous or cancerous enlargements, or swelling of the glands, and allied structures in Peyer's patches, solitary follicles, the spleen and the tonsils. I have found an increase of white cells in cases of cancer, when the cachexia has been very profound.

This condition differs from leucocythæmia then in its transitory character, by the absence of deficiency of the red corpuscles, and by the degree in which the white corpuscles are diminished.

Von Jaksch and Tumas maintain that leucocytosis occurs regularly in croupous pneumonia.

Microcythæmia.—Various observers have from time to time described bodies in the blood which appear like small red corpuscles. Manassein ("Centralblatt," 1871) produced changes in the diameter of the red blood cells by introducing chemical substances into the circulation. He noticed a diminution in size of the corpuscles in cases of hyperpyrexia, in septic poisoning, and when the oxidation of the blood was interfered with, as during the inhalation of carbonic acid. He found the diameter of the corpuscles increased by the action of alcohol and quinine, also under the influence of cold, in cases where an excess of oxygen had produced an acute anæmia.

The bodies known as "microcytes" were first described by Vaulair and Masius ("Bulletin de l'Académie Royale de Méd. de Belgique," 1871, p. 515). They are spherical in form, rather redder than the ordinary corpuscles, and have a diameter of 2 or 3 μ. They do not run into rouleaux. Some authors (Gram, Hayem) consider that they are caused by the fluids added to the blood for diluting purposes, but they are seen when no such agents have been used, and when the pure blood has been examined directly. Portions of colouring matter may be removed by the said agents, and simulate closely the microcytes, but they will be found not to possess such clear outlines, and as a rule only occur sparsely, whilst the microcytes are usually numerous.

They are found in various conditions, such as infectious diseases, pronounced anæmia, and burns. Vaulair and Masius met with them in pathological states in which there was a diminished activity of the

z 2

liver, with an increased activity on the part of the
spleen. I have also observed a similar case. In pernicious anæmia they are very abundant ; Eichorst goes
so far as to consider them pathognomonic of this disease.
As regards their nature, some authorities believe that
they are caused by the rapid and uniform abstraction of
water from the corpuscles, others that they are the final
stage of the changes which the red corpuscles undergo
in the spleen, and that they are destroyed by the liver.
Their true significance is not yet known, and the matter
requires further elaboration before any diagnostic importance can be attached to them.

Poikilocytosis (fig. 75).—In certain affections many

FIG. 75.—Blood in poikilocytosis.

of the red corpuscles are found to have undergone remarkable changes in shape and size. Some are seen
to be much larger than usual (megaloblasts), but the
majority show various departures from the normal
spherical form. Some are oval, others pear-shaped,
others kidney-shaped, whilst some exhibit minute processes, like small knobs projecting from them. Von
Jaksch (*loc. cit.*, p. 24) considers that in this condition
the red corpuscles are endowed with an abnormal power
of contractility.

These appearances were at one time thought to be
peculiar to pernicious anæmia, but have since been
observed in many cases in which the blood is greatly

affected, such as severe anæmia of whatever kind, and in chlorosis. They have also been recorded as occurring in the blood in cancerous cachexia, and in myeloid degeneration of the viscera. The most pronounced specimen I have ever seen was obtained from a patient suffering from gout and profound anæmia with splenic enlargement.

To observe these bodies the blood should be examined pure, without the addition of diluting fluids.

Melanæmia.—In the blood of patients who have suffered for some time from malarial disease, particles of pigment are observed in the blood. These are usually black, more rarely brown or yellow. They are sometimes free, and sometimes enclosed in round or oval cells, probably leucocytes.

This appearance is often associated with an enlarged and deeply pigmented condition of the spleen ; it is not improbable that the pigment particles are produced by the malarial fever, and find their way from the spleen into the blood.

PARASITES IN THE BLOOD.

Vegetable parasites.—Several of the most important pathogenic bacteria have been found in the blood of patients suffering from the particular diseases caused by the germs. The number of microorganisms, however, falls far short of that found in animals under similar conditions.

In order to search for the organisms, the finger of the patient should be well scrubbed with soap and water, and then washed over with corrosive sublimate of the strength of 1 in 1000. The tip of the finger is then compressed and pricked with a sterilised needle. The

drop of blood which exudes is taken off with a clean
cover-glass and immediately covered with another, the
two being pressed together and then slid apart, so as to
obtain two films; when dry they are passed three times
through the flame, and afterwards stained in an appro-
priate manner, according to which bacillus is to be
sought for.

Bacillus of anthrax.—Cover-glasses prepared as
above should be stained by Gram's method (see p. 95).
Their appearance does not differ from that described on
p. 305, as obtained from the local lesion.

Bacillus of tubercle.—Cover-glass preparations
must be stained by Neelsen's method (p. 98). Tubercle
bacilli are found with difficulty in the blood, and a large
numbers of specimens must be obtained. Meisels
found them in the blood of a patient suffering from
miliary tuberculosis and his discovery has since been
confirmed by other observers.

Although I have made several attempts, I have never
been able to detect the bacilli in the blood of tubercular patients.

Bacillus of typhoid fever.—Some observers
have succeeded in finding bacilli in the blood of
patients afflicted with enteric fever, which morphologically, in their manner of growth on artificial media, and
in the effect they have when inoculated on animals, are
similar to those which are found in typhoid stools and
in the tissues after death (see p. 227). Cover-glass preparations should be stained in methylene blue, and
washed in water acidulated with a drop or two of acetic
acid.

Bacillus of glanders.—The pathogenic bacilli
of glanders, in addition to being found in the farcy buds,
and discharges from ulcers in patients suffering from

that complaint, have been found in the blood in small numbers. They are difficult to stain ; cover-glass preparations are best coloured by Kühne's method (see p. 92).

Bacillus of tetanus.—Nicolaier and Kitasato were the first to describe this organism. The rods are long and slender and exhibit terminal spore formation. They have only very rarely been found in the blood.

Spirillum of relapsing fever.—The organism found in the blood of patients suffering from relapsing fever may be demonstrated in a drop of fresh blood placed under a cover-glass, or dried preparations may be made in the usual way, stained with fuchsine, and washed in water containing a few drops of alcohol.

The spirillum is usually known as the "spirillum Obermeieri," having been originally described by Dr. Obermeier of Berlin in 1873. It takes the form of a delicate, homogeneous, spirally-twisted filament, having a length of from $\frac{1}{2000}$ to $\frac{1}{500}$ inch (15-40 μ). In fresh blood it exhibits active movements, rotating on its long axis, and with a lashing movement progressing backwards and forwards.

The "spirochætæ," as the spirilla are sometimes termed, are only found in the blood during the actual attacks of relapsing fever, and cannot be found during the intervals. According to Heydenreich, they appear before the febrile attack begins, but cease to be discoverable before the commencement of the crisis. The same observer also finds that their numbers vary greatly from day to day. Sometimes, after the filaments have been present for two or three days, they suddenly disappear, but a few hours later may reappear in large numbers.

They are very sensitive to reagents of all kinds. If

the blood be examined in the intervals of the disease, peculiar refractive bodies resembling diplococci are found ; these are especially numerous when the paroxysm commences. Von Jaksch (*l.c.*, p. 31) considers that when the attack begins, these bodies grow out, as it were, into short thick rods, from which the spirilla are finally evolved. They are therefore probably spores.

Hæmatozoa of malaria.—Lavaran in 1880 was the first to notice the existence of peculiar structures in the blood of patients suffering from malaria, and his observations have been confirmed by several observers, notably by Marchiafava and Celli.

The different forms met with may be divided into two groups ; those within the red blood corpuscles and those free in the serum.

Intra-corpuscular bodies.—There are three varieties of these :—Firstly, irregular protoplasmic bodies much smaller than the corpuscles. They exhibit active amœboid movements. Marchiafava and Celli who were the first to observe them suggested the name Plasmodium malariæ. Secondly, minute masses of finely granular or hyaline protoplasm enclosing granules of pigment. They are usually circular in shape, and likewise exhibit amœboid movements. The pigment granules also change their shape. Thirdly, forms which appear like isolated grains, and in addition large homogeneous bodies surrounded by clear spaces.

Extra-corpuscular bodies.—These are also of various forms. Firstly, what are known as the "semi-lunar bodies of Laveran." These objects are crescent-shaped, either rounded or pointed at the extremities ; occasionally they are almost spherical. They are motionless. Pigment granules are found in the centre of these bodies, which seem to be composed of homogeneous

protoplasm.   Secondly,   protoplasmic   masses,   finely
granular in appearance.   The pigment is collected into
the form of a rosette, and the protoplasm gradually
undergoes segmentation, forming small spherical bodies,
which are ultimately set free.   Thirdly, circular pear-
shaped, or ovoid bodies, a little smaller than red blood
corpuscles, provided with one or more actively mobile,
long lash-like filaments.   Fourthly, small spherical pigmented bodies about one quarter the size of the red
blood corpuscles.   These exhibit amœboid movements.

To demonstrate these bodies fresh blood is examined
in the usual way.   For permanent specimens, dried
cover-glasses are prepared, and stained by the Gram
method.

Klebs and Tommasi-Crudeli found bacilli in the soil
of the Campagna, which they regard as the specific
organism of malaria.

ANIMAL PARASITES.

These are the Filaria sanguinis hominis and the
Bilharzia hæmatobia.

Filaria sanguinis hominis (fig. 76).—This parasite has already been referred to (see p. 190).   The discovery of the embryos of this nematode worm in the
blood was first made by Dr. T. R. Lewis in 1872.
Four years later the parent worm was found by Dr.
Bancroft of Brisbane.   It has been shown to inhabit
the lymphatics of the inhabitants of tropical countries.
It does not necessarily produce disease and many
natives (1 in 10 in South China) in perfect health were
found to be affected.   Usually the condition is brought
to notice by the patients becoming afflicted with chy-

luria or elephantiasis. Dr. Stephen Mackenzie found them in the blood of a native of the Upper Congo suffering from " sleeping sickness." The parent worm is about 3 to 3½ inches long, and has a diameter of about $\frac{1}{100}$ inch, and resembles "a delicate thread of catgut, animated and wriggling " (Manson). The mouth is circular, without papillæ; there is a narrow neck, and the tail is bluntly pointed. It is with the embryos, however, that we are immediately interested. Specimens are obtained by pricking two or three fingers of the patient, transferring the drops to a glass and covering with a long cover-glass.

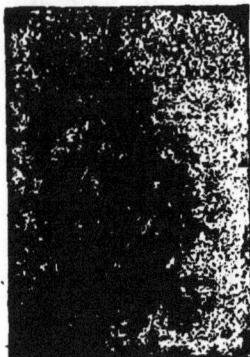

FIG. 76.—Filaria sanguinis hominis (specimen in the urine, Boyce).

The embryos are seen as active organisms about $\frac{1}{90}$ inch in length, and having a diameter $\frac{1}{3200}$; each is enclosed in a delicate sac or sheath which fits it accurately, except that a collapsed or unoccupied part is seen projecting at one end. Most observers agree that this sheath is the envelope or shell of the ovum, which, as the embryo developes becomes stretched out over the skin. This enveloping membrane is quite structureless, but the contained parasite is seen under a high power to be transversely striated and very, granular. The embryos only occur in the peripheral vessels whilst the patient is asleep, generally, therefore, at night. They begin to make their appearance at about 6 or 8 p.m. By midnight their numbers reach the maximum. As morning approaches they become fewer and fewer, and finally disappear altogether; but that this is not due simply to daylight, was shown by Dr. Stephen Mackenzie, who caused a patient to be up all night, and in bed all day, with the result that the filariæ were only found during the day; on the other hand, the most dense London fog, if the patient was up and about during the day, did not tempt the parasites to appear.

Dr. Manson has twice obtained ova in a much earlier stage of development, consisting of oval bodies $\frac{1}{500}$ inch in length, and $\frac{1}{750}$ in breadth.

FIG. 77.—Bilharzia hæmatobia, male and female; eggs (von Jaksch).

Bilharzia hæmatobia (distoma hæmatobium). Patients who have dwelt in tropical countries, and become affected with endemic hæmaturia, have been shown to be the hosts of a parasite, to which the name of Bilharzia hæmatobia was given by Dr. Cobbold. The adult parasites (fig. 77) occupy smooth walled spaces in communication with the veins of the pelvis. This fluke is white in colour, the male and female being distinct. The former is flattened, and about half an inch in length; the hinder part is more cylindrical in form, from the edges being thinned, and folded inwards so as to form a groove in which the female is received during sexual congress. The female is round and thin, and about ¾ inch long. The parasite itself is but rarely seen in the blood, but the ova are discharged, and are often found in the urine (see p. 190).

EXAMINATION OF BLOOD STAINS.

A medical man is occasionally called upon to give a decision as to whether a stain on some portion of clothing is caused by blood. Although in important cases such a question is generally referred to an expert, yet all doctors ought to be prepared to make such an examination. The suspected stain, should first be examined by a good lens; if caused by blood, it will not be a mere colouring of the fibres, but will have a shiny glossy appearance, and each fibre will be invested with a portion of dried coagulum or clot.

The most simple plan of procedure, when small particles of coagulum can be removed, is to place them on a glass slide, and then breathe over them several times. A cover-glass is applied, and if blood be present, a red margin will soon appear, and by the aid of a quarter-inch lens the red blood corpuscles may be recognised. Should this be unsuccessful, a very small drop of water should be run in under the glass—any excess must be avoided, or the corpuscles will swell and loose their characteristic appearances.

Such a simple method, however, is not always practicable. The stained portion of the material must then be cut out and macerated in a small quantity of equal parts of glycerine and water, or better in a mixture proposed by Hofmann—

Water . . . . 300 parts
Glycerine . . . . 100 ,,
Chloride of sodium   . . . .   2 ,,
Corrosive sublimate   . . . .   1 part.

If small portions of the fibres of the material be placed under the microscope, corpuscles will be seen either free in the fluid or entangled in the meshes of the fibres.

A very important question often arises, granted that the stain is caused by blood, is it human? At present our knowledge on this subject is very limited. In all animals with red blood the globules have a disc-like form. In the mammalia with the exception of the camel tribe the outline of the corpuscle is circular. In this tribe, and in birds, fishes and reptiles, they are oval in form. In the three last mentioned classes of animals, they possess a central nucleus. We are able to say with safety then that blood corpuscles are either mammalian and may be human, or that being oval in shape, the corpuscles came from one of the other classes above mentioned.

Dr. Lionel Beale in his work on " The Microscope in Medicine" figures a large number of corpuscles, which are worthy of reference.

CHAPTER XXII.

Cutaneous Parasites.

In order to complete the account of the most important parasites found in disease, a short description must be given of those producing various affections of the skin.

The clinical features are usually quite sufficient upon which to base a diagnosis, but the evidence afforded by the microscope will corroborate the opinion, and in a few atypical cases may yield the only means of arriving at a definite conclusion.

Some of the bacteria producing cutaneous affections, such as those pathognomonic of tubercle, leprosy, &c., have already been considered. .

Cutaneous parasites may be divided into animal and vegetable. The latter will be considered first as being the more important.

Vegetable Parasites.

Trichophyton tonsurans.—This is the characteristic fungus of "ringworm." The diagnosis is not difficult clinically, the presence of short, broken-off hairs, in small patches, sufficiently indicating the nature of the complaint.

As corroborative evidence one or two of the hairs should be removed with forceps and soaked for a little time in dilute liquor potassæ. The hair seems to be covered near its roots with an asbestos-like covering, of a dull white colour. After soaking, they are placed on glass slides with a little of the potash solution, a cover-glass applied, and light pressure made so as to separate the component fibres of the hairs.

Attention should first be directed to the hair itself. The surface will be seen to be rough, the cortex and

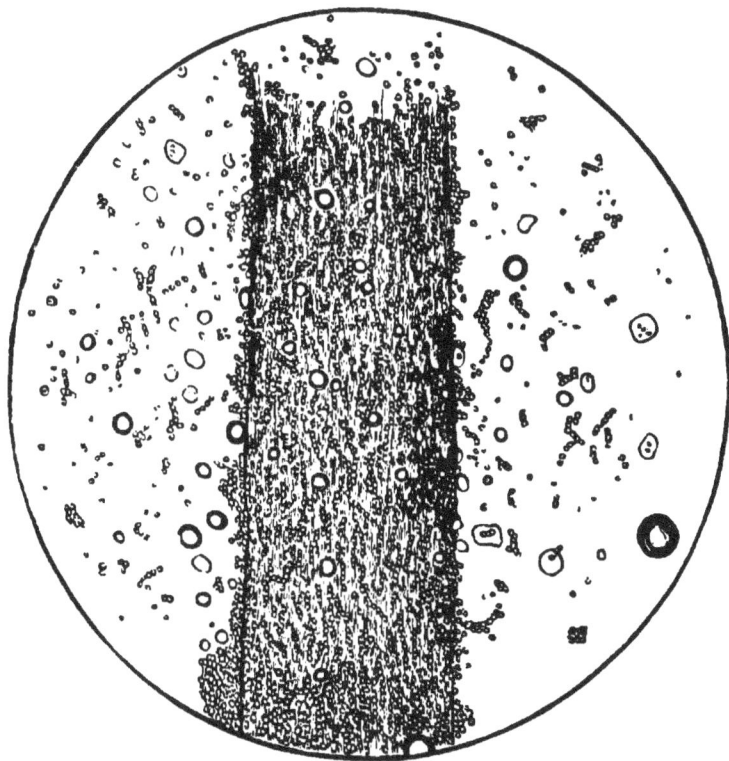

FIG. 78.—Trichophyton tonsurans (Alder Smith).

medulla being almost undistinguishable. The free end, instead of being pointed, is ragged and split. The fungus (fig. 78) consists almost entirely of spores ; these are circular in shape, and about ·004 mm. in diameter; they

possess nuclei, and are chiefly arranged in chains, distributed between the disordered fragments of the hair,
being particularly numerous about the root, the my-

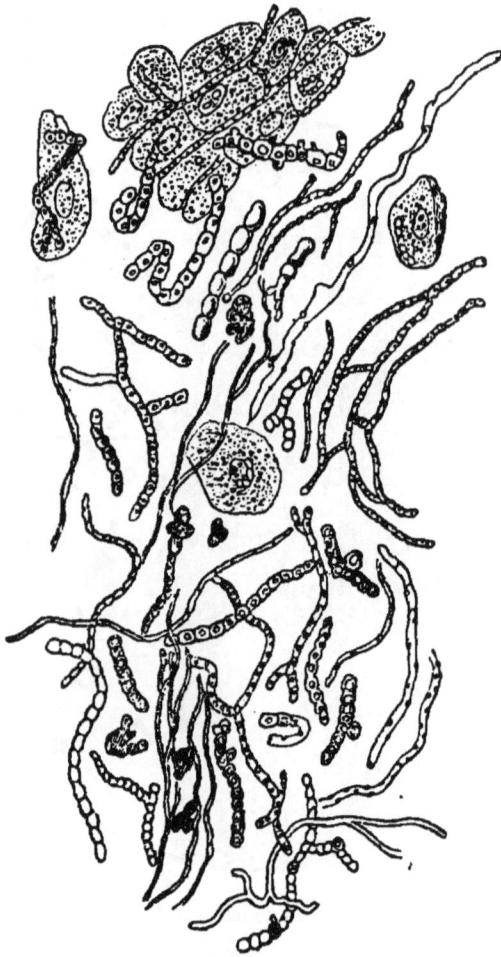

Fig. 79.—Achorion Schönleinii (Crocker).

celial threads are few, jointed, somewhat curved in
their course, and small granules are seen in their interior. When attacking the head this disease is known

as Tinea tonsurans; when the body is affected it is termed Tinea circinata or Tinea marginata.

Tinea marginata, formerly called "eczema marginatum," occurs only in adult males; the same fungus is found in all the forms.

Microsporon furfur (Tinea versicolor).—The affection caused by this fungus is also known as Pityriasis versicolor. It occurs as yellowish-brown spots, scarcely rising above the level of the skin, and distributed generally over the body, especially over the chest and back.

To demonstrate the fungus some of the scales should be removed and examined in a drop of dilute potash. If an opportunity should occur of obtaining sections of skin thus affected, very satisfactory specimens may be obtained by staining them with Weigert's modification of the Gram method (see p. 95). The spores are found collected together into heaps, and are a little larger than those of Tricophyton tonsurans. The mycelial threads are very numerous and form a thick wavy mass, between the component parts of which the spores can be seen.

Achorion Schönleinii.—If a small piece of the crust from a case of favus (fig. 79) is examined in liquor potassæ, dense masses of mycelia will be seen ; they are often so thick that the spores can scarcely be made out. These latter are oval or circular in shape and a little smaller than those of Tricophyton tonsurans.

ANIMAL PARASITES.

Pediculi.—Three species of lice are parasitic on man ; (1) Pediculus capitis; (2) Pediculus vestimenti vel corporis ; (3) Pediculus pubis.

AA

1. **Pediculus capitis.**—In all cases of pustular inflammation of the scalp the hair should be carefully searched for pediculi, although the impetigo which results is produced more by the scratching of the patient than by the irritation of the lice.

Children are most frequently attacked, and women more than men.

The louse is about a line in length, of a dirty-white colour, and covered with short scattered hairs. The abdomen is oval, and distinct from the head and thorax. The head has two short antennæ, and large prominent black eyes. The parasite has six well-developed legs, furnished with strong claws. The male, which is rather smaller than the female, has a peculiar long projection on its back—the penis. The animal itself cannot always be found, but equally diagnostic are the presence of "nits." These contain the ova, and are small, white, semi-transparent bodies, made of a hard material and triangular in form. They are firmly attached to the hairs, about three quarters of an inch from the roots, by short pedicles being glued to them by a material secreted from the lice.

2. **Pediculus vestimenti.**—In general appearance this louse closely resembles the pediculus capitis ; it is, however, rather larger in size. Its ova are deposited in the clothing and not on the surface of the body.

3. **Pediculus pubis.**—As its name indicates this parasite chiefly attacks the hair about the generative organs. It is different in shape to the other two varieties, resembling a small crab. It is also smaller, and the line of separation between the abdomen and thorax is not so well marked. The legs are short and curved, and terminated by strong claws.

Sarcoptes hominis.—This parasite, also known

as the Acarus scabiei, or itch mite, produces the well known disease scabies. The clinical features are too well known to need description here, and we need only consider the parasite itself.

The male insects are but rarely seen, as they do not burrow but live upon the surface of the body. The female, after impregnation, makes her way under the skin, forming the characteristic cuniculus or "run." To demonstrate the mite, a burrow should be laid open with a stout needle, from the entrance to its blind extremity; the acarus will then be seen as a small white particle, and usually clings to the point of the needle. It should be transferred to a glass slide and covered with liquor potassæ, or if it be desired to preserve it, the best medium for the purpose is glycerine jelly.

The animal is in shape something like a tortoise, it has four pairs of legs, and is covered with a chitinous integument furnished with short spines and scattered hairs. The female is larger than the male and on the four anterior legs suckers will be noticed. The male has suckers on all the legs.

Another method for more completely examining the mite and its burrow is to excise the burrow, parasite and all, by means of a sharp pair of scissors, although this is rather a difficult little operation. By means of a good lens, the passage can then be seen, often containing a row of oval eggs in chitinous cells, with the female acarus lying at one end.

Acarus folliculorum.—This is a small parasite occasionally found in the hair follicles and sebaceous glands. It cannot be said to be pathological as it is found in about ten per cent. of all healthy adults.

The acarus may be demonstrated by squeezing out the contents of some of the follicles and placing them on a glass slide with dilute liquor potassæ.   The animal will then be seen to possess a pointed abdomen, distinct from the head and thorax which are continuous. It has four pairs of legs each terminating in three strong claws.

CHAPTER XXIII.

EXAMINATION OF FOOD AND WATER.

THE examination of drinking water is almost entirely conducted by chemical analysis, although an investigation of the sediment by the microscope, and especially by bacteriological processes, may yield important information. The adulterations of food are detected partly by chemical and partly by microscopical methods. We shall naturally consider here only the points that can be determined by the use of the microscope; these are so many that only the most important can be touched upon. If the student is desirous of going more deeply into this subject, he cannot do better than consult the work by Dr. A. H. Hassall, " Food : its Adulterations, and the Methods for their Detection."

EXAMINATION OF FOOD.

In looking down the list of substances commonly used for the adulteration of food, the various cereals occupy a prominent place, and an intimate knowledge of their structure and the appearances of their grains is absolutely necessary for the purposes under consideration, and we shall therefore begin with a description of these bodies, afterwards indicating the food-matters with which they may be found mixed.

It may be here stated, however, that food is adulterated with three main objects, firstly, to increase its

bulk and weight ; secondly, to improve its appearance
and colour ; thirdly, to add to its taste, smell or other
properties.

Wheat.—Several structures enter into the formation
of the grain of wheat, as well as that of the other cereals.

The seed is surrounded by membranes, technically
known as the "testa"; the surface of the seed consists of angular cells, filled with granular oily matter,
whilst its substance is made up of cells filled with
starch corpuscles, and it is with these latter that we are
most closely concerned, as the testa is in great part re-

FIG. 80.—Wheat-starch grains.

moved in the process of grinding and dressing the flour,
and the same may be said of the cells forming the surface of the grain.

The general structure of a " starch grain " is as follows :—Each granule exhibits a peculiar spot, termed
the "hilum," round which are seen a set of circular
lines which are for the most part concentric with it.   A
characteristic effect is produced by polarised light, each
grain shows a dark cross, the point of intersection being
at the hilum.

The starch grains of wheat (in fact of all cereals) are
best viewed with a quarter-inch lens.   They will be

seen to consist of round or oval particles of various sizes
(fig. 80), some being very small and others of considerable size, whilst only a few are of intermediate dimensions. The small grains are chiefly circular in shape,
with a central hilum ; the larger granules form rounded
or flattened discs, with thin edges, in these the hilum
and concentric lines are barely visible, if at all.

Occasionally some of the larger granules are twisted
or turned up at the edges, and when seen sideways,
present the appearance of a longitudinal furrow, giving
a very good representation of a hilum ; but if the cover-
glass be slightly moved so as cause the grains to turn
over, their real nature is at once apparent.

FIG. 81.—Barley-starch grains.

Portions of the outer envelopes of the wheat grain
may be detected in the coarser and more branny flours.
The testa consists of three layers of cells, two of which
are disposed longitudinally and the third transversely
to the axis of the seeds.

The cells forming the surface of the seeds are large
and angular.

Barley.—Like that of wheat, barley starch (fig. 81)
consists of small and large grains, in fact the two starches
are almost indistinguishable from one another. The
larger grains are as a rule more distinctly ringed, while
a greater proportion of them présent the longitudinal
furrow. If portions of the testa can be secured the

difference from wheat is seen to be more marked. The testa of the barley grain consists of four layers of cells, smaller than those of wheat. There are three layers of longitudinal cells which are not beaded as in the case of wheat.

The cells of the surface of the grain are not nearly so large as those of wheat, and there are three layers instead of one.

Rye.—In general form the starch grains resemble those of wheat, but a few differences may be noticed. The lesser grains are decidedly smaller than the corresponding grains of wheat, and many of the large rye granules have a peculiar rayed hilum. With the polariscope

FIG. 82.—Oatmeal-starch grains.

they exhibit a very strongly marked cross, thus distinguishing them from barley-starch.

The testa of rye also closely resembles that of wheat.

Oatmeal (fig. 82).—The starch grains are smaller than those of wheat and more uniform in size; they are polygonal in shape and no hilum or concentric rings are visible. They tend to cohere together forming rounded masses, presenting a reticulated surface. With polarised light, unlike the starches of other cereals, no crosses are produced. The longitudinal cells of the testa are large and well-defined, and long and pointed hairs arise from some of them. The transverse cells form a single layer, and are almost square in shape.

Maize (Indian corn flour).—The starch grains (fig. 83) are polygonal in outline, and exhibit well-marked central depressions, and occasionally a radiated hilum. They are larger in size than those of the oat, and do not tend to cohere into masses. With the polariscope a well-defined cross is visible.

FIG. 83.—Maize-starch grains.

The testa is made up of two membranes, the outer consisting of seven or eight layers of cells which are much longer than they are broad, and the margins of the outermost layer are beaded. The inner membrane consists of a single layer of cells.

The cells containing the starch granules are very angular and are sub-divided by numerous septa.

FIG. 84.—Rice-starch grains.

Rice (fig. 84).—In shape the starch grains resemble those of the oat, being polygonal ; they are, however, much smaller. They exhibit well-marked central depressions and raised edges.

The testa is rather complex. The outer surface is

thrown up into ridges, arranged both longitudinally and transversely.   The ridges contain silica in the form of granules.   The substance of the husk is composed of short fibres, arranged like the ridges.   Beneath this fibrous membrane is a thin layer of angular cells, which are longer than they are broad.

Beans (fig. 85).*—Bean starch cells are more or less oval and somewhat flattened ; they exhibit a longitudinal cleft in the centre of the grain, crossed by transverse fissures.

Peas (fig. 86).—Pea starch grains are smaller than

FIG. 85.—Bean-starch cells.   FIG. 86.—Pea-starch grains.

those of the bean and less flattened, but otherwise similar in shape.   The longitudinal cleft which runs the whole length of the grain does not show the transverse fissures.

Sago (fig. 87).—The starch is obtained from the pith of several kinds of palms.   The starch grains are elongated in form and of considerable size ; they are larger at one end than at the other.   The large end is rounded, and the other truncated.   The hilum is usually circular, but sometimes cracked, when it appears as a

* Figures 80-94 are borrowed with the kind permission of the author from Parkes' " Hygiene and Public Health."

slit or star. In some of the granules a few concentric rings may be seen.

Tapioca (fig. 88).—The starch is procured from the root of the plant. The granules are small in size, sometimes united into groups of three or four. They are roughly circular in form, although owing to mutual pressure they may become truncated at one extremity as with the sago grains. The hilum is round and well-marked.

Arrowroot (fig. 89).—There are several kinds of arrowroot. The starch grains of each variety differ slightly from each other, but it would be of no advan-

FIG. 87.—Sago-starch grains.   FIG. 88.—Tapioca-starch grains.

tage to describe each form separately. Generally speaking the grains are oval or pyriform in shape; the corpuscles of "British arrowroot" are often circular. Their size varies greatly, but they are usually large. The hilum is situated in the broad extremity of each granule, and the concentric lines are fine, regular, and although crowded together, are well marked.

Potato (fig. 90).—The starch grains of potato are very characteristic. They are of large size. Most of them are pyriform in shape, others almost circular. The hilum is situated at the narrow end. The striæ are well marked, but are not so regular or so numerous as in arrowroot starch, and the lines appear to be coarser.

Turmeric.—This substance is easily recognised by its bright yellow colour. When crushed, in addition to almost structureless débris, starch granules, pyriform in shape with well-defined hilum and concentric lines are seen.

This completes a description of the various substances used in the adulteration of the chief articles of food, and we have now to indicate the principal articles of diet which are adulterated by them.

FIG. 89.—Arrowroot-starch grains.　FIG. 90.—Potato-starch grains.

Arrowroot.—The best kind of arrowroot is generally considered to be the Maranta or West Indian. The starch grains are usually oval in form, but sometimes almost triangular ; they vary greatly in size. The hilum is situated at the broad end of the grain whilst in the inferior kinds of arrowroot it is found at the narrow end. It is not always circular, but is frequently seen as a short sharp line running transversely. In the other kinds of arrowroot the hilum is nearly always circular, so that by the form and position of the hilum maranta arrowroot can be readily distinguished from the starches used to adulterate it.

In order to increase its bulk and weight, arrowroot is mixed with "sago, potato, and tapioca starches, and various mixtures and combinations of these with the inferior arrowroots " (Hassall). With a little practice, all these substances can be recognised with the micro-scope.

Bread.—Bread should be made of pure wheat flour, but many inferior forms are often added in order to increase its bulk and weight. The most common adul-terations are mashed potatoes, rye, maize, rice and beans. Alum and salt are also often added in large quantities, but they can only be detected by chemical analysis.

Flour.—Wheat flour is chiefly adulterated with rice, beans, maize, rye, and potato-flour. The microscope forms the only means of recognising these impurities. Alum and other inorganic salts are sometimes added in large quantities.

Wheat and other cereals are liable to become in-fected, and consequently deteriorated, by the develop-ment of various parasites, animal and vegetable.

Smut (Uredo segetum).—This disease forms yellow spots upon the stem, leaf and chaff. When a scraping is made, placed under the microscope and examined under a quarter-inch lens, minute sporules are seen, of granu-lar aspect and many of them presenting a double con-tour.

Bunt (Uredo fœtida).—The spores are larger than those of "smut," and exhibit a reticulated appearance. This fungus only attacks the grains of wheat.

Rust (Puccinia graminis).—The spores form yellow or brownish oval spots upon the stem and leaf. The spores are intermediate in size, between those of "rust" and "smut." In the earliest stage of growth they are round, but afterwards become oval, and attached by a short slender stalk to the surface on which they develop; after a time they become free.

Ergot (Oïdium arbortifaciens).—When ergot affects the grain, the seed coat and starch cells are replaced by a layer of dark cells, the large cells by the small cells of the ergot, and the starch grains by drops of oil. A bloom appears on the grain, and consists of the sporidia of the fungus. If wheat be badly stored or allowed to become damp, various moulds and fungi form, such as *mucor mucedo, penicillium glaucum* and *aspergillus niger.*

Ear-cockle (Vibrio tritici).—This parasite fills the grain,

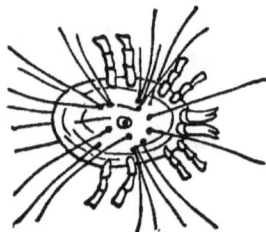

FIG. 91.—Acarus farinæ.

which becomes green and then black, with a cotton-like substance. Under the microscope this material is seen to be composed of numerous slender animalculæ, which when placed in water exhibit the most active movements.

In addition are frequently found the *Acarus farinæ* (fig. 91), and the weevil or Calandra granaria, a small insect, visible to the naked eye, which eats the core out of the grain leaving only the shell.

Oatmeal.—The characters of the starch grains have been described on page 360. " It would hardly be supposed that sufficient inducement exists for the sophistication of an article like oatmeal; it appears, however,

that the supposition is not correct. . . . Of thirty
samples of oatmeal submitted to examination, sixteen,
or rather more than one-half, were found to be adulterated with large quantities of barley meal " (Hassall).
Barley costs about half as much as oats, hence the
inducement.   .

The investing membranes of the oat, barley and
wheat, technically termed " rubble" or " sharps " are
also added.

The testa of the various cereals have been described
in the early part of the chapter, and it is mainly on
their recognition by the microscope that the detection
of the adulteration depends.   Rice and maize are sometimes employed in addition to the cereals named above.

BEVERAGES.

Tea.—The structure of the tea leaf is characteristic,
although it varies somewhat according to age.   It is
oval in shape and has a serrated border ; the primary
veins run out alternately from the mid-rib, and turn
towards the point of the leaf, but without reaching the
border.

The chief adulteration formerly consisted in mixing
other leaves with those of the tea plant, but since the
duty has been greatly reduced and the Customs' authorities have their own appointed analysts, such adulteration is now almost unknown.   Mineral matters are
sometimes largely employed, such as sand, quartz and
magnetic oxide of iron.

For the detection of these adulterations chemical
means are usually employed, but it is also necessary to
have a thorough knowledge of the structure of the tea
leaf, and of the chief leaves that may be mixed with it.

368   MEDICAL MICROSCOPY.

Extensive practice alone can give the necessary information.

Coffee.—We are concerned only with the structure
of the coffee berry. The testa or inverting membrane
is chiefly made up of elongated oval cells (fig. 92) ad-
herent to one another and presenting irregular cross
markings on their surfaces; they form a single layer
and rest upon a thin membrane having an indistinct
fibrous structure.

The substance of the berry consists of an irregular

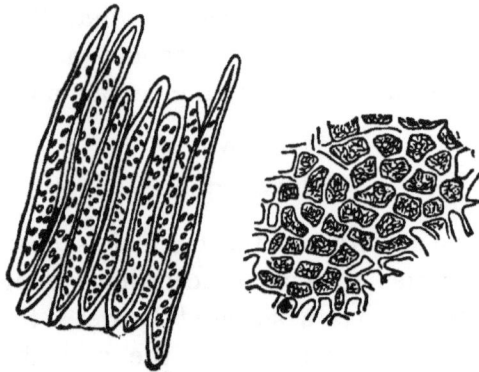

FIG. 92.—Coffee; cells of testa and cellular structure.

network of fibres forming a cellular structure, in which
is contained a considerable quantity of oily matter.

The most prevalent adulterant of coffee is chicory.

Chicory is the root of a plant not the seed. It can be
fairly readily distinguished from coffee under the microscope. The ducts have a strong resemblance to the
cells of the testa of coffee, but they present a dotted
surface (fig. 93) which sufficiently distinguishes them.
The chief part of the root is made up of cells embedded
in a coarse areolar tissue. The accompanying figures
show the main differences between the two structures.

Coffee in addition is adulterated with roasted wheat, rye and potato flour, roasted beans, mangel wurzel and acorns, and to improve its colour, burnt sugar is sometimes employed.

Cocoa.—The starch grains of cocoa are very small

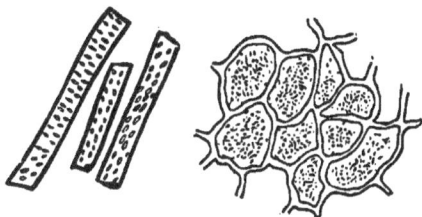

FIG. 93.—Chicory. Dotted ducts and cellular structure.

and circular in form (fig. 94). When the seed is compressed it breaks up into pieces, which are known as "nibs," and within the cells forming these, the starch grains are found.

The chief adulterants of cocoa are various forms of starches, namely Maranta, East India and Tahita

FIG. 94.—Cocoa-starch cells.

arrowroot; the flours of wheat, maize, sago, potato and tapioca ; sugar, chicory and cocoa husks. All these substances may be recognised by the microscope.

To improve its colour, Venetian red, red ochre and other ferruginous earths are added.

EXAMINATION OF DRINKING WATER.

As already stated chemical analysis is by far the most important means of examining drinking water, but not infrequently the nature of the suspended matters have to be determined and this is accomplished with the microscope.

When the water is turbid, no difficulty will be found in obtaining a sediment, but as a rule the deposit is slight.

Dr. J. D. Macdonald ("A Guide to the Microscopical Examination of Drinking Water") gives the following directions for the collection of specimens :—"A tall glass vessel is taken, and filled up to the neck with the water to be tested.   A circular disc of glass resting upon a horizontal loop at the end of a long wire is let down to the bottom of the vessel, which is then covered and set aside for twenty-four or forty-eight hours.   At the end of this period the water is siphoned off with a piece of india-rubber tubing, leaving only a thin stratum over the glass disc.   This is now carefully raised and laid upon a piece of blotting-paper, so as to dry its under surface, when it may at once be transferred to the stage of the microscope, with a large piece of thin covering glass so placed as to exclude all air-bubbles."

Another good plan is not to siphon off all the water, but to leave a little, then shake the deposit up with it, and pour the mixture into a tall conical glass, from which portions of the sediment may be taken up by means of a pipette and examined in the usual way.

Various mineral substances and dead vegetable and

animal matter will then be observed, together with many living forms both animal and vegetable.

We cannot enter here into a description of the numerous objects which present themselves in various samples of water, for to do so would entail several additional chapters. The reader who is interested in such matters cannot do better than consult Dr. J. D. Macdonald's book already alluded to, where he will find a full description of a large number of organisms and a series of beautifully executed lithographic plates.

Bacteriological methods are also now largely employed in the examination of drinking water. The samples are collected in sterilised flasks plugged with cotton-wool.

The apparatus for plate cultivations (see p. 389) must be prepared, shallow cultivation dishes being very convenient.

A tube of liquefied nutrient gelatine is taken, and the plugs of both it and the flask having been removed, a drop of water is transferred to the tube by means of a sterilised graduated pipette, the plugs being immediately replaced. The tube is rotated so as to diffuse the organisms through the gelatine, and a plate cultivation is then made. When the gelatine has set, the plate is transferred to a damp chamber. Colonies will commence to develop in two or three days and may then be examined.

" Drop cultivations " may also be made in sterile bouillon.

With plate cultivations the number of colonies which develop are usually counted, so as to form some estimate of the number of micro-organisms in a measured quantity of the water under examination. For this purpose a glass plate, ruled by horizontal and vertical lines into centimetre squares, some of which are again divided into ninths, is so arranged on a wooden frame that it can cover the plate without touching it. The colonies in a certain number of squares are counted, so that the number of colonies on the entire surface can be calculated by a process of multiplication.

Cover-glass preparations should be made from the colonies and stained in the usual way.

CHAPTER XXIV.

BACTERIOLOGICAL METHODS.

BEFORE a micro-organism can be considered to be patho-genic it has to satisfy certain conditions. These have been laid down by Koch as follows :—

1. The micro-organism must be found in the blood, or other fluids or tissues of the animal suffering from the disease in question, and in no other disease.

2. The micro-organisms must be isolated and culti-vated on nutrient media outside the body.

3. A pure cultivation thus obtained must produce the same disease when introduced into the body of a healthy animal.

4. The same micro-organisms must be found in the animal inoculated.

In order to carry out the investigations necessary for demonstrating the above postulates, certain apparatus is required. Bacteriology is essentially an exact science, and unless every single detail is carefully adhered to in the various processes adopted, failure can be the only result. The apparatus that is absolutely needed is more complicated than that required for the ordinary pathological methods which have been described in the previous chapters. A good gas and water supply must be at hand, and for any really satisfactory work a room should be set apart for the purpose, in other words, a well-appointed laboratory is almost indispensable.

It is not proposed in this chapter to enter deeply into

the methods of bacteriological research, for this purpose
a large volume would be needed, but only such proced-
ures and apparatus will be described, by which micro-
organisms can be isolated and cultivated in the simplest
way.

It is difficult to detail on paper the delicate manipula-
tions which have to be learnt, and I should strongly
advise the student who wishes to study the science of
bacteriology, wherein lies a huge field for investigation,
to commence by taking out a " course" of lessons at one
of the laboratories now open for the purpose.

In order to carry out original investigation a sound
knowledge of the rudiments of the science is necessary,
and this can only be obtained by practical study and
experience.

APPARATUS REQUIRED.

Steam steriliser.
Hot-air steriliser.
Hot-water filter.
Incubator.
Test-tubes.
Glass flasks.
Large cultivation dishes.
Shallow cultivation dishes.
Platinum needles.   Scalpels.
Funnels ; filter-paper, &c.

Steam steriliser (fig. 95).*—This consists of a
cylindrical vessel, made of leaded iron with a copper
bottom.   It is about two feet in height, and is jacketed

* Figures 95-101 are borrowed with the kind permission of the
author from Prof. Crookshank's " Manual of Bacteriology."

with thick felt.   The lid is conical, has handles on either
side, and is perforated at the apex to hold a thermometer.   On one side is a gauge which marks the level
of the water, and a tap for running it off.   Inside the
vessel is an iron grating about two-thirds of the way
down which separates the water from the steam chamber.
The apparatus stands upon three legs, and is heated

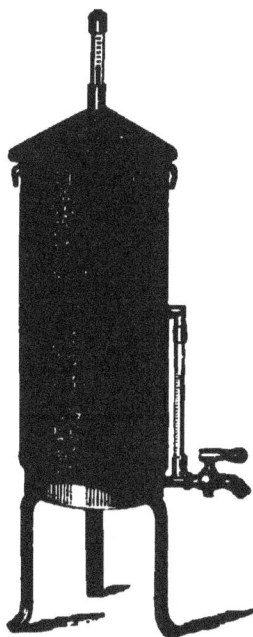

FIG. 95.—Koch's steam steriliser.

from below with a Fletcher's burner, or two or three
Bunsen's.   When the water commences to boil, the lid
is removed and the flasks or tubes with their contents,
which have to be sterilised, are placed within it, resting
on the grating.   Test-tubes are usually arranged in a
wire basket, so that a large number may be stacked
together.

Hot-air steriliser.—This is a square box, made of sheet iron, with double walls ; it may either be hung against the wall, or be supported on four legs. In the roof is a hole for the thermometer, and it is heated from below by a Fletcher's burner. The temperature can be

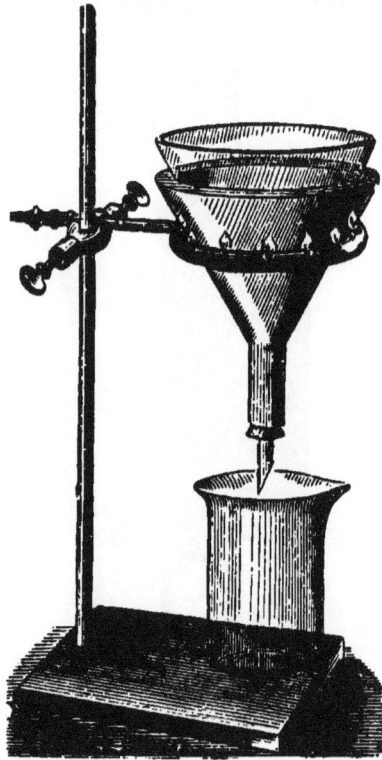

FIG. 96.—Hot-water filtering apparatus with ring burner.

raised to 150° C. Cotton-wool, test-tubes, flasks, &c., may be sterilised in it, being kept at a high temperature for about an hour. Cotton-wool is most conveniently enclosed in glass capsules.

Hot-water filter (fig. 96).—This apparatus consists of a copper funnel with double walls, the space between them being filled with water, which is kept hot by means of a circular burner, which at the same time serves as a funnel ring.

Within the copper cone fits a glass funnel which is fixed by means of caoutchouc plug. It is employed for filtering the solid cultivating media which are liquid when hot.

Another arrangement is to have a metal receiver containing the funnel placed in connection with the steam steriliser, which thus serves a double purpose.

Incubator.—Most incubators are rectangular chests with double walls, the interspace being filled with water and kept at a constant temperature by means of a " gas regulator." There are several varieties, but the one known as " Babes' " is about the best (fig. 97). The sides and roof are covered with felt. In the roof are two hooks, one for a thermometer and the other for the gas regulator. The front of the chest is provided with a glass door and a sliding glass panel, which are likewise closed in by a sheet of felt. It is heated from below by two protected burners which are supplied with gas through a " thermo-regulator."

The best form of regulator is Reichert's, a full description of which will be found in the text-books on Bacteriology.

Test-tubes.—A large number will be required, the most convenient size is about six inches long by five-eighths of an inch in diameter.

Glass flasks.—Large flasks are employed for boiling nutrient media, and small conical ones for storing solid cultivating material.

Large cultivating dishes.—These are chiefly used for potato cultivations, also for "plate cultivations."

They consist of two parts, the larger and upper fitting over the smaller and lower.

Small cultivating dishes.—These are much smaller and shallower than the preceding, they should be procured as flat as possible, as they are employed instead of plates for isolating colonies of bacteria.

FIG. 97.—Babes' incubator.

Platinum needles (fig. 98).—These are easily made with stout platinum wire and glass rods. The wires—about two or three inches in length—are forced into the end of the rods by heating the ends of the latter until soft enough to admit them. At least three will be required: one having a loop at its extremity, another being bent into a short hook, whilst the other is allowed to remain straight.

Potato knife.—This consists of a broad smooth bladed knife set in a wooden handle.

The other accessory apparatus need not be described, and may be procured as required.

In all matters connected with the microscope, *cleanliness* cannot be too strongly insisted upon, but with bacteriological work it is, if possible, even more important, for without it, it is impossible to proceed. All apparatus and instruments must not only be clean to the

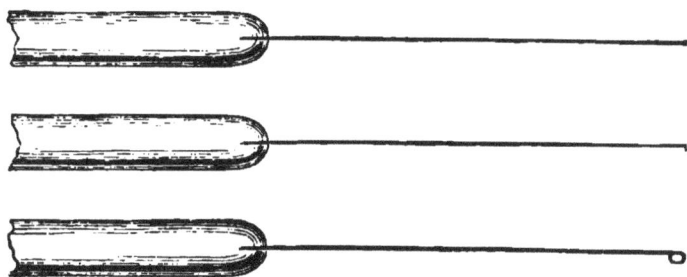

FIG. 98.—Platinum needles ; straight, hooked, looped.

naked eye, but must be thoroughly purified from all germs.

This is accomplished mainly by heat. The word that must be constantly dinned into the ear of the beginner is "*sterilise.*" Metal instruments may simply be heated to redness in a Bunsen burner. The scalpels must afterwards be placed with their handles on the table, and their blades projecting over the edge. It should become a routine practice always to heat the platinum wires both before and after use.

Cotton-wool should be sterilised by being placed in the hot-air steriliser.

Glass apparatus, test-tubes, flasks, &c., after being

thoroughly washed should be swilled out with distilled water, next with corrosive sublimate solution (1 in 1000), then rinsed with alcohol and ether, after which they are plugged with cotton-wool and placed in the hot-air steriliser for an hour, when they may be considered to be ready for use.

CULTIVATING MEDIA.

A very large number of cultivating media have been introduced, but only those most commonly used will be described here. They are of two kinds, solid and liquid.

SOLID MEDIA.

Potato cultivations.—This is the most simple form of cultivating medium.

The usual way of conducting the process is as follows :—

Smooth-skinned potatoes should be chosen as free as possible from " eyes " and rotten spots, should any of these be present they must be picked out with the point of a knife. Each potato is scrubbed with a hard brush and then soaked for half an hour in a five per cent. solution of corrosive sublimate.

They are next placed in a " potato receiver " and heated in the steam steriliser at a temperature of 100° for about half an hour, after which they are allowed to cool.

Meanwhile, the " damp chambers " or large cultivation dishes are prepared in the following manner. The glass is cleaned and washed in corrosive sublimate solution as above described. A sheet of filter paper is cut to the size of the receiver, and after being dipped in the corrosive sublimate is placed in the bottom of the dish, the cover being at once replaced. The potato knife and two or three scalpels are also sterilised.

The operator should then wash his hands and swill them with sublimate solution.

When the potatoes are cool, an assistant raises the top of the potato receiver, and a potato is picked out and held between the fingers and thumb of the left hand. By a clean cut with the large knife, the potato is almost divided into two. The assistant now raises the upper part of the cultivating dish, and the potato is quickly placed on the damp paper, a final turn of the knife separating the two parts, so that they lie side by side, cut surface uppermost ; the dish must be immediately covered. A second potato is then treated in the same way, all four portions being left in the same receiver.

After again cleansing the hands, one portion of potato is removed from the dish, and a small portion of the substance to be inoculated placed on the centre with a sterilised platinum needle. With one of the prepared scalpels, the material is distributed over the surface of the potato, a clear margin of about quarter of an inch being left. The potato is replaced and another half taken up, the receiver being closed in the intervals. With another scalpel a small portion is removed from the inoculated surface of the first potato and spread over the surface of the second, and the process is repeated from the second to the third, and from the third to the fourth, so procuring " first," " second " and " third " attenuations. The inoculations are then left to grow.

Another method of potato cultivation, which is now much employed is to cut the potato after cooking into small squares (of course with a sterilised knife), and place them in large test-tubes, and inoculate them with a platinum wire, or to cut small cubes and keep them in small covered capsules, which have been cleansed and sterilised.

Nutrient gelatine.—Half a kilogramme (one pound) of fresh beef is taken, as free from fat as possible, cut into small pieces, passed through a mincing machine, and placed in a large flask. A litre of distilled water is added, the whole well shaken and placed in a refrigerator or ice-pail for twenty-four hours. It is then strained through linen, and the red juice thus obtained placed in a measuring glass, and made up to a litre by the addition of more distilled water. To the resulting liquid, contained in a large flask, are added:— 10 grammes of dried peptone, 5 grammes of common salt, and 100 grammes of best gelatine. The mixture is gently heated in a water bath, and in a short time the gelatine will be dissolved. The flask should be occasionally shaken.

When this part of the process is complete, the fluid must be neutralised or rendered slightly alkaline. For this purpose several pieces of red litmus paper should be laid on the table, and a concentrated solution of carbonate of soda added to the mixture, drop by drop, a glass rod being frequently dipped into the flask, and then one of the pieces of litmus touched with it until the red paper becomes faintly blue. If too much alkali has been added, the error may be rectified with a little lactic acid.

The liquid is next heated in a water bath for an hour, its reaction being tested once or twice during that period. Afterwards it is filtered through paper, the hot water filter being used for the purpose. A pale straw-coloured filtrate should result ; unless perfectly clear, it must again be passed through the filter.

Whilst still hot it is poured into test-tubes and flasks (which have been cleaned and sterilised in the manner already described). A sterilised funnel should be employed for the test-tubes, so as to prevent the mixture touching the upper part of the tubes, which should be filled for about a third of the depth, and the cotton-wool plugs immediately replaced.

When filled, the tubes are placed in their basket, and, with the flasks, heated in the steam steriliser for ten minutes, this being repeated on two successive days. On the third day, whilst still hot, some of the tubes should be inclined, so that the medium will solidify diagonally along part of the tube, instead of vertically.

The tubes and flasks are now ready for inoculation.

Nutrient agar-agar.—This preparation remains solid up to a temperature of about 45° C. The first part of its preparation is conducted in the same way as was described in the preparation of nutrient gelatine, that is to say, the meat is minced, allowed to stand in distilled water for twenty-four hours, and then filtered. Instead of adding gelatine, however, twenty grammes of agar-agar are added to the litre of meat infusion. The solid substance should be cut up into small pieces with scissors, and then dropped into the flask. After boiling for an hour, the mixture is filtered through flannel, the hot water apparatus being again employed. It will probably be necessary to filter twice, and even then it is difficult to obtain a perfectly clear fluid ; but if only a slight cloudiness appears, it is fit for use. A small quantity of liquid collects in the tubes after a time, this is no disadvantage, but care must be taken not to allow it to touch the cotton-wool plugs. Both "straight" and "oblique" tubes should be prepared. The tubes must be sterilised by being placed in the steam steriliser for forty-five minutes on three successive days.

Glycerine agar-agar.—This medium is particularly used for cultivating tubercle bacilli. It is prepared in the same way as agar-agar, with the exception that after the mixture has been boiled, five per cent. of glycerine is added, and the whole is then filtered.

Sterilised blood serum.—Koch's method of preparing this medium is as follows :—Two or three glass cylinders are cleansed and sterilised in the way already described. They should be fitted with glass stoppers, which ought to be greased with vaseline. The part of the animal's skin from which the blood is to be obtained is shaved, and thoroughly cleansed by washing with corrosive sublimate solution and a mixture of alcohol and ether. With a sterilised scalpel a blood vessel is opened, and after the first jet of blood has escaped, the remainder is allowed to flow directly into the cylinders, which when full, and the stoppers replaced, are deposited in a refrigerator or ice pail for twenty-four hours. By that time the clots will have separated, and the clear amber-coloured serum is drawn off in sterilised pipettes, and placed in sterilised test-tubes, which are plugged with cotton-wool. The tubes should be filled about a third of their length, and they are then put into a steriliser, at a temperature of 58°, for about two hours, and this process repeated on three successive days. The serum has now to be solidified, and for this purpose the tubes are arranged on a specially constructed apparatus, consisting of a tin vessel with double walls, between which hot water is kept, and covered on the top with a plate of glass. When laid on it the tubes are inclined so that the fluid solidifies obliquely. The temperature is raised to between 65° and 68° C., and the tubes carefully watched. Directly solidification takes place they must be removed, and the medium should appear transparent and of a pale straw colour.

LIQUID MEDIA.

The liquid cultivating media are not so much employed for bacteriological research as the solid media. They have serious disadvantages; in the first place, many pathogenic organisms will not grow well in them, and secondly, it is difficult to be sure that the cultivation is a pure one. Much may, however, be learnt from their use, especially in "drop-cultures," and in some inoculation experiments on animals. Such a medium is also largely employed, when it is desired to undertake a chemical examination of the culture, and by passing it through a porcelain filter to separate the bacteria from the cultivating fluid, and so obtain only their products.

Bouillon.—Beef, pork, or chicken may be the basis from which the bouillon is made. The meat is cut up and minced in the same way as for nutrient gelatine, mixed with distilled water in the same proportions, the infusion allowed to stand in ice for twenty-four hours, filtered through linen, and the filtrate brought up to the litre with more distilled water; 10 grammes of dried peptone and 5 of common salt are then added. The mixture is neutralised with carbonate of soda in the way already described, and the flask of broth deposited in the steam steriliser at a temperature of 100° C. for half an hour. The liquid is afterwards filtered, and the

clear filtrate transferred to sterilised test-tubes, which
are plugged, and the medium sterilised by being placed
in the steam steriliser for half an hour on three successive days.

Liquid blood serum.—This is prepared in the
way already described, except that the final process for
coagulation of the serum is omitted. A few artificial
media are occasionally employed, of which the two
following are the most important :—

Pasteur's fluid.—This liquid has the following
composition :—

Pure cane sugar . . . . . 10 parts
Ash of yeast . . . . . . ·075 part
Ammonium tartrate. . . . 1 part
Distilled water . . . . . 100 parts.

Cohn's fluid.—This has been modified by Mayer
and consists of :—

Phosphate of potassium . . . ·1 part
Sulphate of magnesium . . . ·1 ,,
Ammonium tartrate . . . . ·2 ,,
Tribasic calcium phosphate . ·01 ,,
Distilled water . . . . . . 20·0 parts.

MODE OF USING THE SOLID NUTRIENT MEDIA.

Test-tube cultivations.—The tubes should be
kept in the dark until wanted, and then placed in a
beaker, in the bottom of which a small quantity of
cotton-wool has been placed.

The cultivations are of two kinds, according to
whether the gelatine has solidified obliquely or vertically. The former tubes are employed for surface cultivations, whilst the latter are used for " deep growths."

The mode of making the "deep growths" will be described first.

The cotton-wool plug is twisted out, so as to loosen any adhesions that may accidentally have formed, and held between the fourth and fifth fingers of the left hand. The tube is held mouth downwards in the same hand (fig. 99). A straight platinum needle which has recently been sterilised and allowed to cool, is dipped

FIG. 99.—Method of inoculating a test-tube containing sterile nutrient jelly.

into the fluid containing the bacteria, or a colony is removed from a plate cultivation, and the wire introduced carefully into the tube without touching the sides, and plunged once well into the jelly, quickly withdrawn, and the tube immediately plugged. Half a dozen tubes should be thus inoculated, and then put away in the dark and examined daily to observe the growth of the cultivation.

Surface growths.—These are made in two ways:—In the first, the plug is removed as before, and the sterilised straight platinum wire is drawn once along the surface, after having been dipped into the fluid to be tested, or having on it a colony from a plate, or the results of a potato cultivation, &c., and the plug rapidly replaced.

In the second method the substance to be inoculated is diluted so that the formation of colonies may be observed.

About a drachm of normal saline solution is boiled in a sterilised test-tube, which is immediately plugged with sterilised cotton-wool and allowed to cool. A small particle of the pure culture or whatever material is under experiment is then introduced on a platinum wire and mixed with the solution by rotating the tube. A gelatine tube is now taken and the two held side by side as in fig. 100. The two plugs are removed and held one between the fourth and fifth, and the other between the third and fourth fingers of the left hand. A sterilised platinum wire is then dipped into the saline solution and drawn once across the surface of the gelatine, and afterwards the plug is replaced. Several gelatine tubes should be inoculated from the same saline solution.

A few additional details may be here added. It is rather inconvenient to hold the cotton-wool plugs as above directed, and the following procedure may be adopted :—

A piece of gauze wire is heated to redness and placed on a tripod stand. The plugs are removed with sterilised forceps instead of with the fingers, and whilst the tubes are being inoculated are laid on the gauze wire. When they are again removed, they are placed in the flame of the Bunsen burner, so that their external surfaces are scorched, the flame being extinguished by blowing it out, or better by the mere fact of their being placed back in the tubes.

The plugs should be pushed down the tube until only a small piece projects, this should be again set fire to, the flame extinguished by blowing and the mouths of the tubes covered with caoutchouc caps made for the purpose.

The tubes should be labeled with the nature of their contents and the date of inoculation. If the culture has been taken from another tube, " second generation " or " first sub-culture," &c., should be added.

Surface inoculations on agar-agar tubes and on blood serum are made in the same manner.

Plate cultivations.—Test-tube cultivations are best adapted for the growth of bacteria in pure cultures. In order to procure such cultures, " plate cultivations " have to be made in the following manner :—The glass plates are similar to those used by photographers. They must be thoroughly cleansed, swilled in corrosive sublimate solution, then in alcohol, and finally placed in iron boxes and heated in the hot-air steriliser for several hours. Some tubes of nutrient gelatine are placed in a beaker containing warm water, so as to liquefy the gelatine, and the cotton-wool plugs loosened so that they can be taken out without delay.

One of the tubes is now inoculated with some of the fungoid material, by means of a sterilised platinum wire ; the needle should be rotated a few times so as to mix the bacteria thoroughly with the gelatine. Another tube is now taken and the cotton-wool removed, the two tubes being held side by side (fig. 100). The plugs may either be laid between the fingers or laid on sterilised wire-gauze as already described. A sterilised platinum wire terminating in a loop is now passed care-

fully into the first tube, and a drop removed and mixed
with the gelatine in the second tube.   The original tube
is now plugged, and a third one opened and a drop
removed from the second and placed in it ; the operation being repeated with a fourth tube.   The tubes
must now be labeled thus :—first tube, original ; second
tube, 1st dilution ; third tube, 2nd dilution; fourth tube,
3rd dilution.   The cotton-wool plugs should be tem-

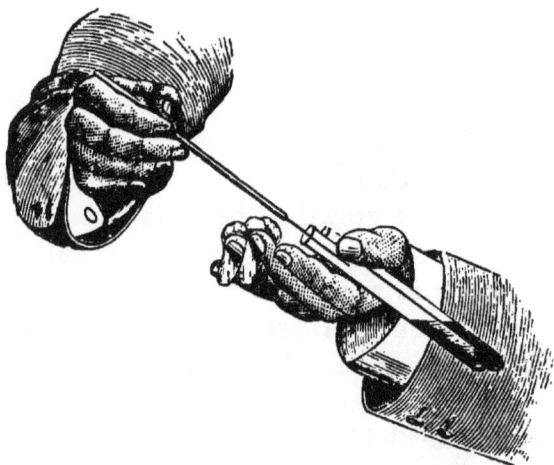

FIG. 100.—Method of inoculating test-tubes in the preparation of
plate cultivations.

porarily replaced, with the necessary precautions, and
the tubes placed in a beaker.

The leveling apparatus is now made ready.   This
consists of a tripod stand with screw legs, so that by
placing a spirit level on it, it can be made perfectly
level.   On the stand is placed a large glass dish containing iced water, and covered with a sheet of glass
which has been sterilised.   The glass apparatus may
be replaced by a sheet of iron which is rather to be preferred ; in the winter cooling may be dispensed with,

but in the summer it must be laid on ice immediately before use. Damp chambers must also be prepared as described on p. 380.

One of the glass plates (which have in the meantime been allowed to cool) is now removed from the box with sterilised forceps and laid on the glass or iron plate, and then covered with a bell-jar.

The " original " tube is now taken and a part of the rim heated in the Bunsen burner. When nearly cool, the plug is removed and laid aside. With the left hand the bell-jar is raised, and the contents of the tube poured on to the plate, the gelatine being distributed as evenly as possible within a short distance of the edge of

FIG. 101.—Damp chamber for plate cultivations.

the plate, by means of the part of the lip of the test-tube which has been heated. The bell-jar is replaced until the gelatine has solidified, when the plate is removed to the " damp chamber " (fig. 101) and laid on a " bridge "; this consists of a glass plate about six inches long and two wide, supported at each end by two glass bars about half an inch thick. A similar bridge is laid over the plate resting on the lower support.

Plates are prepared in a similar manner from the other tubes, a careful memorandum being kept of the order in which they were placed, or labels may be attached to the bridges.

The cover must of course be replaced as each of the plates is deposited. When the series is complete the glass dish is set aside. In a short period, varying according to the nature of the bacteria, small colonies will commence to grow, gradually increasing in size and assuming their characteristic appearance.

If the original culture was pure, only one form of colony will be seen, but if not several varieties may be seen. On the "original" plate they will probably be so numerous as not to be separable, but in the various dilutions they will become more and more discreet, so that in the third and perhaps also in the second, the individual colonies will be clearly separated from one another.

One of the plates is placed on the stage of the microscope and examined with a low power. The colonies will appear more sharply defined if a small diaphragm is used. A characteristic colony is selected and arranged so that it occupies the centre of the field.

A sterilised platinum wire bent at the extremity into a small hook is introduced between the object glass and the plate. When the hook is seen close to the selected colony, the needle is dipped into it, raised and withdrawn, carrying the whole or part of the colony with it, and a gelatine or agar-agar tube is immediately inoculated with it, or a cover-glass preparation may be made of it for the purpose of ascertaining the nature of the microbe. The above process is a little difficult at first, but with some practice it can be accomplished with rapidity and precision, and a pure culture assured. There are two modifications of " plate cultivations."

Shallow cultivation dishes, carefully cleaned and sterilised, may be used instead of plates, so avoiding the use of the leveling apparatus. One dish must of course correspond to each test-tube, and the covers se-

cured by means of rubber bands. This mode of procedure is simpler than the plates, and is equally as
efficacious. The "dilutions" of the culture must of
course be made in the same way as above described.

Esmarch introduced a method which entirely does away
with plates or dishes. After the test-tubes have been
inoculated, each one is covered with an india-rubber
cap over the plug of cotton-wool. It is then held almost
horizontally under a running cold water tap, the upper
end being slightly raised. The tube is now rotated on
its long axis, and the gelatine thus solidifies round the
sides of the tube.

The colonies, when formed, are easily examined, both
with a high and a low power, and may be removed by
means of a platinum wire for examination or further
cultivation.

Plate cultivations of agar-agar tubes are made in the
same way, the plates being afterwards placed in an
incubator.

MODE OF USING THE LIQUID MEDIA.

Tubes containing liquid media are inoculated simply
by removing the plug, and introducing a sterilised platinum needle carrying the organisms.

Drop cultures.—These are very important for
studying the life-history of the organisms, and for
watching their movements.

Glass slides excavated in the centre are required.
To make a culture, one of the slides is cleaned, and
then sterilised by being held, depression downwards, in
the flame of a Bunsen burner. A little vaseline is
painted round the excavation, and the slide covered
with a large basin or bell-jar.

A cover-glass is next cleaned and sterilised. A drop of a liquid culture is then placed on it with a platinum looped needle, or a drop of sterile bouillon is deposited on it and inoculated from some culture. The prepared glass slide is now held over it, so that the drop comes exactly in the centre of the depression. Slight pressure is used so that the cover-glass adheres to the vaseline. The preparation is turned over and a rim of vaseline painted round the glass. The culture can then be examined from day to day with an oil-immersion lens, using a diaphragm below the Abbé's condenser.

EXAMINATION OF MICRO-ORGANISMS.

Cover-glass preparations are made thus :—A clean cover-glass is taken and a drop of distilled water placed on it. A sterilised platinum wire is introduced into a culture tube, and a minute portion of the growth removed, or a colony may be removed from a plate cultivation. The point of the needle is lowered into the drop of water, which is distributed over the cover-glass, carrying the micro-organisms with it. When dry the glass is passed three times through the flame of a Bunsen burner, and then stained according to the nature of the microbe which is being examined.

Cover-glass impressions.—These are employed to demonstrate the minute structure of colonies on plate cultivations.

A clean sterilised cover-glass is dropped on to a plate cultivation, a part being preferably selected where there are several typical well formed colonies. It is gently pressed down, and then one side is raised with a sterilised needle. The glass is next removed with sterilised forceps, allowed to dry, passed three times through the flame of a Bunsen burner and stained.

The methods employed for staining the principle pathogenic micro-organisms have been described in the previous chapters. For the best way of colouring those not mentioned, textbooks on bacteriology must be consulted.

Experiments on animals.—The usual method of infecting animals is by subcutaneous inoculation.

A small patch of hair is shaved from a part of the body which the animal cannot easily reach with its tongue, and the skin washed with corrosive sublimate solution (1 in 1000). A small incision is made with a sterilised scalpel, and a portion of the cultivation placed in the wound, which is allowed to close over it. In guinea-pigs and rabbits the inside of the upper part of the hind leg is the part usually selected.

More usually the inoculation is made in a fluid form. A watch glass is cleaned and sterilised by heat. A little normal saline solution which has been boiled for one minute is poured into it, and some of the material to be inoculated mixed with the fluid. A freshly prepared capillary pipette is taken, the point broken off with sterilised forceps, and some of the liquid drawn up into it, the amount varying with the nature of the substance to be experimented on. A small incision is made in the skin of the animal, prepared in the way already described. In guinea-pigs a spot just internal to the mamma is usually chosen. The extremity of the pipette is introduced into the subcutaneous cellular tissue and the fluid blown out of the tube. The animal must be firmly held by an assistant.

Another method of inoculating fluids is by means of a syringe. Koch's instrument is the most convenient,

as it can be so easily washed and sterilised.   It consist
of a graduated glass cylinder, to the lower extremity c
which a hollow needle is attached and to the upper en
an india-rubber bellows provided with a stop-cocl
Before and after use it must be washed out with di:
tilled water which has been boiled for three minute:
then with absolute alcohol and finally with more of th
distilled water.

The needle is introduced into the watch glass cor
taining the mixed culture and normal saline solutio
(or a cultuie in a liquid medium) and drawn up into th
syringe by squeezing the balloon and loosening it agair
finally turning the tap.   The injection is made into th
subcutaneous tissue or peritoneal cavity.

Another mode of infecting animals is by feedinɡ
For this purpose, the animals should be fed, for at lea:
a month beforehand, on food which has been rendere
sterile by boiling, and the pure cultivation or other ir
fective material is then intimately mixed with th
sterile food.   This method is only applicable to certai
micro-organisms, as most bacteria are destroyed by th
action of acids of the stomach.

The animal may also be placed in a closed chambe:
and the germs introduced into the atmosphere by mean
of a spray-producer, the cultivation being mixed wit
sterilised water.

All animals chosen for experiments should be a
healthy as possible.   They must be well attended t(
their cages carefully cleaned daily and suitable foo
must be given.   They should be kept under observatio
at least a month before they are inoculated so as to e>
clude as far as possible the possibility of previous ir
fection.

WORKS CONSULTED.

Only the more important works are included here, other references are incorporated with the text.

BEALE, LIONEL, " How to work with the Microscope."
　　,,　　,,　　" Microscope in Medicine."
BIZZOZERO ET FIRKET, " Microscopie Clinique."
BRISTOWE, " Textbook of Medicine."
CARPENTER, " The Microscope."
COATES, " Manual of Pathology."
COBBOLD, " Entozoa."
COLMAN, " Section Cutting and Staining."
CROOKSHANK, " Manual of Bacteriology."
DAVIS, G. S., " Practical Microscopy."
FAGGE, " Principles and Practice of Medicine."
FRÄNKEL, " Pathologie und Therapie der Krankheiten des Respirations Apparates."
FRIEDLÄNDER, " Microscopische Technik."
GIBBES, HENEAGE, " Practical Physiology and Pathology."
GREEN, " Textbook of Pathology."
HAMILTON, " Textbook of Pathology."
HARRIS AND POWER, " Physiological Laboratory."
HASSALL, " Food and its Adulteration."
VON JAKSCH, " Klinische Diagnostik " (English translation by Cagney).
" Journal of Microscopical Society."
KÜHNE, " Practical Guide to the Demonstration of Bacteria in Animal Tissues " (Translation by Dr. Vincent Harris).
LEE, " Microtomist's Vade Mecum."

Leuckart, "Parasites of Man."

Macdonald, J. I., "Guide to the Microscopical Examination Drinking Water."

Mackenzie, Hunter, "Examination of Sputum."

Morris, "Diseases of the Kidney."

Moullin, Mansell, "Surgery."

Nothnagel, "Berträge zur Physiologie und Pathologie des Darmɛ

Orth, "Compendium der Diagnostik."

Parkes, "Hygiene and Public Health."

Pflüger's "Archiv."

Powell, Douglas, "Diseases of the Lungs and Pleuræ."

Quain, "Dictionary of Medicine."

Ralfe, "Diseases of the Kidney."

Roberts, F. T., "Textbook of Medicine."

Roberts, Sir W., "Diseases of the Kidney."

Schäfer, "Practical Histology."

Stirling, "Outlines of Practical Histology."

"Transactions of Pathological Society."

"Transactions of Royal Medical and Chirurgical Society."

Troupe, "Examination of Sputum."

Ultzmann, "Impotence and Sterility."

Virchow, "Archiv."

Walsh, "Diseases of the Lungs."

Woodhead, "Practical Pathology."

Wynter and Wethered, "Manual oı Clinical and Practi Pathology."

Ziegler, "Pathologische Anatomie."

INDEX.

SELECTED LIST
OF
NEW AND RECENT WORKS
PUBLISHED BY
H. K. LEWIS,
136 GOWER STREET, LONDON, W.C.
(ESTABLISHED 1844)

₊ *For full list of works in Medicine and Surgery published by*
H. K. Lewis see complete Catalogue sent post free on application.

WILLIAM ROSE, B.S., M.B. LOND., F.R.C.S.
Professor of Surgery to King's College, London, and Surgeon to King's
College Hospital, &c.

ON HARELIP AND CLEFT PALATE. Demy 8vo,
with Illustrations. [*Just ready.*

F. CHARLES LARKIN, F.R.C.S. ENG.
Late Assistant Lecturer in Physiology in University College, Liverpool,
AND
RANDLE LEIGH, M.B., B.SC. LOND.
Senior Demonstrator of Physiology in University College, Liverpool.

OUTLINES OF PRACTICAL PHYSIOLOGICAL
CHEMISTRY. Second Edition, With Illustrations,
crown 8vo, 2s. 6d. *nett.* [*Just ready.*

DR. THEODOR PUSCHMANN.
Public Professor in Ordinary at the University of Vienna.

HISTORY OF MEDICAL EDUCATION FROM THE
MOST REMOTE TO THE MOST RECENT TIMES.
Translated by EVAN H. HARE, M.A. (OXON.), F.R.C.S. (ENG.),
F.S.A. Demy 8vo, 21s. [*Just ready.*

SIDNEY COUPLAND, M.D., F.R.C.P.
Physician to the Middlesex Hospital, and Lecturer on Practical Medicine in the
Medical School; Examiner in Medicine at the Examining Board for
England.

NOTES ON THE EXAMINATION OF THE SPU-
TUM, VOMIT, FÆCES, URINE AND BLOOD.
Second Edition, 12mo, 1s. *nett.* [*Just published.*

4000 –11/91.

New and Recent Works published by

SIR WILLIAM AITKEN, KNT., M.D., F.R.S.
Professor of Pathology in the Army Medical School.

ON THE ANIMAL ALKALOIDS, THE PTOMAINES, LEUCOMAINES, AND EXTRACTIVES IN THEIR PATHOLOGICAL RELATIONS. Second edition, crown 8vo, 3s. 6d.

E. CRESSWELL BABER, M.B. LOND.
Surgeon to the Brighton and Sussex Throat and Ear Dispensary.

A GUIDE TO THE EXAMINATION OF THE NOSE WITH REMARKS ON THE DIAGNOSIS OF DIS-EASES OF THE NASAL CAVITIES. With Illustrations, small 8vo, 5s. 6d.

JAMES B. BALL, M.D. (LOND.), M.R.C.P.
Physician to the Department for Diseases of the Throat and Nose, and Senior Assistant Physician, West London Hospital.

A HANDBOOK OF DISEASES OF THE NOSE AND NASO-PHARYNX. With Illustrations, post 8vo, 6s.
[*Now ready.*]

G. GRANVILLE BANTOCK, M.D., F.R.C.S. EDIN.
Surgeon to the Samaritan Free Hospital for Women and Children.

I.
RUPTURE OF THE FEMALE PERINEUM. Second Edition, with Illustrations, 8vo, 3s. 6d.

II.
ON THE USE AND ABUSE OF PESSARIES. With Illustrations, Second Edition, 8vo, 5s.

FANCOURT BARNES, M.D., M.R.C.P.
Physician to the Chelsea Hospital; Obstetric Physician to the Great Northern Hospital, &c.

A GERMAN-ENGLISH DICTIONARY OF WORDS AND TERMS USED IN MEDICINE AND ITS COGNATE SCIENCES. Square 12mo, Roxburgh binding, 9s.

H. CHARLTON BASTIAN, M.A., M.D., F.R.S.
Examiner in Medicine at the Royal College of Physicians; Physician to University College Hospital, etc.

PARALYSES: CEREBRAL, BULBAR, AND SPI-NAL. A Manual of Diagnosis for Students and Practitioners. With numerous Illustrations, 8vo, 12s. 6d.

ALFRED H. CARTER, M.D. LOND.
Fellow of the Royal College of Physicians; Physician to the Queen's
Hospital, Birmingham, &c.

E LEMENTS OF PRACTICAL MEDICINE.   Sixth
Edition, crown 8vo, 9s.   [*Just published.*

FRANCIS HENRY CHAMPNEYS, M.A., M.B. OXON., F.R.C.P.
Physician Accoucheur and Lecturer on Obstetric Medicine at St. Bartholomew's Hospital; Examiner in Obstetric Medicine in the University
of London, &c.

I.

L ECTURES ON PAINFUL MENSTRUATION. THE
HARVEIAN LECTURES, 1890.   Roy. 8vo, 7s. 6d.
[*Just published.*

II.

E XPERIMENTAL RESEARCHES IN ARTIFICIAL
RESPIRATION IN STILLBORN CHILDREN, AND
ALLIED SUBJECTS.   Crown 8vo, 3s. 6d.

W. BRUCE CLARKE, M.A., M.B. OXON., F.R.C.S.
Assistant Surgeon to, and Senior Demonstrator of Anatomy and Operative
Surgery at St. Bartholomew's Hospital, &c.

T HE DIAGNOSIS AND TREATMENT OF DIS-
EASES OF THE KIDNEY AMENABLE TO DIRECT
SURGICAL INTERFERENCE.   Demy 8vo, with Illustrations,
7s. 6d.

WALTER S. COLMAN, M.B. LOND.
Formerly Assistant to the Professor of Pathology in the University of
Edinburgh.

S ECTION CUTTING AND STAINING: A Practical
Guide to the Preparation of Normal and Morbid Histological Specimens.   Illustrations, crown 8vo, 3s.

W. H. CORFIELD, M.A., M.D. OXON.
Professor of Hygiene and Public Health in University College, London.

D WELLING HOUSES: their Sanitary Construction and
Arrangements.   Third Edition, with Illustrations, crown
8vo.   [*In the press.*

EDWARD COTTERELL, M.R.C.S. ENG., L.R.C.P. LOND.
O N SOME COMMON INJURIES TO LIMBS: their
Treatment and After-Treatment including Bone-Setting (so-
called).   Imp. 16mo, with Illustrations, 3s. 6d.

CHARLES CREIGHTON, M.D.

ILLUSTRATIONS OF UNCONSCIOUS MEMORY IN DISEASE, including a Theory of Alteratives. Post 8vo, 6s.

H. RADCLIFFE CROCKER, M.D. LOND., B.S., F.R.C.P.
Physician, Skin Department, University College Hospital.

DISEASES OF THE SKIN: THEIR DESCRIPTION, PATHOLOGY, DIAGNOSIS, AND TREATMENT. With Illustrations, 8vo, 21s.

EDGAR M. CROOKSHANK, M.B. LOND., F.R.M.S.
Professor of Bacteriology, King's College, London.

I.

MANUAL OF BACTERIOLOGY. Third Edition, coloured plates and wood engravings, 8vo, 21s.

II.

HISTORY AND PATHOLOGY OF VACCINATION. 2 vols., royal 8vo, coloured plates, 36s.

HERBERT DAVIES, M.D., F.R.C.P.
Late Consulting Physician to the London Hospital, and formerly Fellow of Queen's College, Cambridge.

THE MECHANISM OF THE CIRCULATION OF THE BLOOD THROUGH ORGANICALLY DIS- EASED HEARTS. Edited by ARTHUR TEMPLER DAVIES, B.A., M.D. Cantab., M.R.C.P. Crown 8vo, 3s. 6d.

HENRY DAVIS, M.R.C.S. ENG.
Teacher and Administrator of Anæsthetics to St. Mary's and the National Dental Hospitals.

GUIDE TO THE ADMINISTRATION OF ANÆS- THETICS. Fcap. 8vo, 2s.

ARTHUR W. EDIS, M.D. LOND., F.R.C.P.
Senior Physician to the Chelsea Hospital for Women; Late Obstetric Physi- cian to the Middlesex Hospital.

STERILITY IN WOMEN: including its Causation and Treatment. With 33 Illustrations, 8vo, 6s. [*Just published.*

AUSTIN FLINT, M.D., LL.D.
Professor of Physiology and Physiological Anatomy at the Bellevue Hospital
Medical College, New York, &c., &c.

A TEXT-BOOK OF HUMAN PHYSIOLOGY. Fourth edition, with 316 illustrations, royal 8vo, 25s.

J. MILNER FOTHERGILL, M.D.
I.

A MANUAL OF DIETETICS. Large 8vo, 10s. 6d.

II.

INDIGESTION, BILIOUSNESS, AND GOUT IN ITS PROTEAN ASPECTS.
PART I.—INDIGESTION AND BILIOUSNESS. Second Edition, post 8vo, 7s. 6d.
PART II.—GOUT IN ITS PROTEAN ASPECTS. Post 8vo, 7s. 6d.

III.

THE TOWN DWELLER: HIS NEEDS AND HIS WANTS. Post 8vo, 3s. 6d.

FORTESCUE FOX, M.D. LOND.
Fellow of the Medical Society of London.

STRATHPEFFER SPA, ITS CLIMATE AND WATERS, with observations, historical, medical, and general, descriptive of the vicinity. Illustrated, cr. 8vo, 2s. 6d. *nett*.

JOHN HENRY GARRETT, M.D.
Licentiate in Sanitary Science and Diplomate in Public Health, Universities
of Durham and Cambridge, &c.

THE ACTION OF WATER ON LEAD: being an inquiry into the cause and mode of the action and its prevention. Crown 8vo, 4s. 6d.

ALFRED W. GERRARD, F.C.S.
Examiner to the Pharmaceutical Society; Teacher of Pharmacy and Demonstrator of Materia Medica to University College Hospital; etc.

ELEMENTS OF MATERIA MEDICA AND PHARMACY. With the New Official Remedies, 1890. Crown 8vo, 8s. 6d.

JOHN GORHAM, M.R.C.S.
Fellow of the Physical Society of Guy's Hospital, etc.

TOOTH EXTRACTION: A manual of the proper mode of extracting teeth. Third edition, fcap. 8vo, 1s. 6d.

GEORGE M. GOULD, A.B., M.D.
Ophthalmic Surgeon to the Philadelphia Hospital, etc.

A NEW MEDICAL DICTIONARY. A compact concise Vocabulary, convenient for reference, based on recent medical literature. Small 8vo, 12s. 6d.

[*Now ready.*

J. B. GRESSWELL, M.R.C.V.S.
Provincial Veterinary Surgeon to the Royal Agricultural Society.

VETERINARY PHARMACOLOGY AND THERA-PEUTICS. Fcap. 8vo, 5s.

DR. JOSEF GRUBER.
Professor of Otology in the mperia Royal University of Vienna, &c.

A TEXT-BOOK OF THE DISEASES OF THE EAR. Translated from the second German edition by special permission of the Author, and edited by EDWARD LAW, M.D., C.M. EDIN., M.R.C.S. ENG., Surgeon to the London Throat Hospital for Diseases of the Throat, Nose and Ear; and by COLEMAN JEWELL, M.B. LOND., M.R.C.S. ENG., late Physician and Pathologist to the London Throat Hospital. With 150 Illustrations, and 70 coloured figures on 2 lithographic plates, royal 8vo, 24s. [*Just published.*

BERKELEY HILL, M.B. LOND., F.R.C.S.
Professor of Clinical Surgery in University College; Surgeon to University College Hospital, and to the Lock Hospital.

THE ESSENTIALS OF BANDAGING. For Managing Fractures and Dislocations; for administering Ether and Chloroform; and for using other Surgical Apparatus. Sixth Edition, with Illustrations, fcap. 8vo, 5s.

BERKELEY HILL, M.B. LOND., F.R.C.S.
Professor of Clinical Surgery in University College.
AND
ARTHUR COOPER, L.R.C.P., M.R.C.S.
Surgeon to the Westminster General Dispensary, &c.

I.

SYPHILIS AND LOCAL CONTAGIOUS DISOR-DERS. Second Edition, entirely re-written, royal 8vo, 18s.

II.

THE STUDENT'S MANUAL OF VENEREAL DIS-EASES. Fourth Edition, post 8vo, 2s. 6d.

PROCTER S. HUTCHINSON, M.R.C.S.
Assistant Surgeon to the Hospital for Diseases of the Throat.

A MANUAL OF DISEASES OF THE NOSE AND THROAT ; including the Nose, Naso-pharynx, Pharynx, and Larynx. With Illustrations, cr. 8vo, 3s. 6d.   [*Now ready.*

NORMAN KERR, M.D., .F.L.S.
President of the Society for the Study of Inebriety ; Consulting Physician, Dalrymple Home for Inebriates, etc.

INEBRIETY: ITS ETIOLOGY, PATHOLOGY, TREATMENT, AND JURISPRUDENCE. Second Edition, crown 8vo, 12s. 6d.   [*Now ready*

J. WICKHAM LEGG, F.R.C.P.
Assistant Physician to Saint Bartholomew's Hospital, and Lecturer on Pathological Anatomy in the Medical School.

A GUIDE TO THE EXAMINATION OF THE URINE ; intended chiefly for Clinical Clerks and Students. Sixth Edition, revised and enlarged, with additional Illustrations, fcap. 8vo, 2s. 6d.

LEWIS'S POCKET CASE BOOK FOR PRACTITIONERS AND STUDENTS. Designed by A. T. BRAND, M.D. Roan, with pencil, 3s. 6d. *nett.*

LEWIS'S POCKET MEDICAL VOCABULARY. Second Edition, 32mo, limp roan, 3s. 6d.   [*Now ready.*

T. R. LEWIS, M.B., F.R.S. ELECT, ETC.
Late Fellow of the Calcutta University ; Surgeon-Major Army Medical Staff.

PHYSIOLOGICAL AND PATHOLOGICAL RESEARCHES. Arranged and edited by SIR WM. AITKEN, M.D., F.R.S., G. E. DOBSON, M.B., F.R.S., and A. E. BROWN, B.Sc. Crown 4to, portrait, 5 maps, 43 plates including 15 chromolithographs, and 67 wood engravings, 30s. *nett.*

WILLIAM THOMPSON LUSK, A.M., M.D.
Professor of Obstetrics and Diseases of Women in the Bellevue Hospital Medical College, &c.

THE SCIENCE AND ART OF MIDWIFERY. Third Edition, revised and enlarged, with numerous Illustrations, 8vo, 18s.

LEWIS'S PRACTICAL SERIES.

These volumes are written by well-known Hospital Physicians and Surgeons recognised as authorities in the subjects of which they treat. They are of a thoroughly Practical nature, and calculated to meet the requirements of the general Practitioner and Student and to present the most recent information in a compact and readable form; the volumes are handsomely got up and issued at low prices, varying with the size of the works.

HYGIENE AND PUBLIC HEALTH. By LOUIS C. PARKES, M.D., D.P.H. Lond. Univ., Assistant Professor of Hygiene, University College, London; Fellow, and Member of the Board of Examiners of the Sanitary Institute; etc. Second Edition, with Illustrations, cr. 8vo., 9s. [*Now ready.*]

MANUAL OF OPHTHALMIC PRACTICE. By C. HIGGENS, F.R.C.S., Ophthalmic Surgeon to Guy's Hospital; Lecturer on Ophthalmology at Guy's Hospital Medical School. Illustrations, crown 8vo, 6s.

A PRACTICAL TEXTBOOK OF THE DISEASES OF WOMEN. By ARTHUR H. N. LEWERS, M.D. Lond., M.R.C.P. Lond., Assistant Obstetric Physician to the London Hospital; Examiner in Midwifery and Diseases of Women to the Society of Apothecaries of London; etc. Third Edition, with Illustrations, crown 8vo, 10s. 6d. [*Now ready.*]

ANÆSTHETICS THEIR USES AND ADMINISTRATION. By DUDLEY W. BUXTON, M.D., B.S., M.R.C.P., Administrator of Anæsthetics to University College Hospital and to the Hospital for Women, Soho Square. Second Edition, crown 8vo. [*Nearly ready.*]

TREATMENT OF DISEASE IN CHILDREN: EMBODYING THE OUTLINES OF DIAGNOSIS AND THE CHIEF PATHOLOGICAL DIFFERENCES BETWEEN CHILDREN AND ADULTS. By ANGEL MONEY, M.D., F.R.C.P., Assistant Physician to the Hospital for Children, Great Ormond Street, and to University College Hospital. Second Edition, crown 8vo, 10s. 6d.

ON FEVERS: THEIR HISTORY, ETIOLOGY, DIAGNOSIS, PROGNOSIS, AND TREATMENT. By ALEXANDER COLLIE, M.D. (Aberdeen), Medical Superintendent of the Eastern Hospitals. Coloured plates, cr. 8vo, 8s. 6d.

HANDBOOK OF DISEASES OF THE EAR FOR THE USE OF STUDENTS AND PRACTITIONERS. By URBAN PRITCHARD, M.D. (Edin.), F.R.C.S. (Eng.), Professor of Aural Surgery at King's College, London; Aural Surgeon to King's College Hospital. Second Edition, with Illustrations, crown 8vo. 5s. [*Now ready.*]

A PRACTICAL TREATISE ON DISEASES OF THE KID- NEYS AND URINARY DERANGEMENTS. By C. H. RALFE, M.A., M.D. Cantab., F.R.C.P. Lond., Assistant Physician to the London Hospital. With Illustrations, crown 8vo, 10s. 6d.

DENTAL SURGERY FOR GENERAL PRACTITIONERS AND STUDENTS OF MEDICINE. By ASHLEY W. BARRETT, M.B. Lond., M.R.C.S., L.D.S., Dental Surgeon to, and Lecturer on Dental Surgery and Pathology in the Medical School of, the London Hospital. Second Edition, with Illustrations, crown 8vo, 3s. 6d.

BODILY DEFORMITIES AND THEIR TREATMENT: A Handbook of Practical Orthopædics. By H. A. REEVES, F.R.C.S. Ed., Senior Assistant Surgeon and Teacher of Practical Surgery at the London Hospital. With numerous Illustrations, crown 8vo, 8s. 6d.

* *Further Volumes will be announced in due course.*

JEFFERY A. MARSTON, M.D., C.B., F.R.C.S., M.R.C.P. LOND.
Surgeon General Medical Staff (Retired).

NOTES ON TYPHOID FEVER: Tropical Life and its Sequelæ. Crown 8vo, 3s. 6d.   [*Now ready.*

EDWARD MARTIN, A.M., M.D.

MINOR SURGERY AND BANDAGING WITH AN APPENDIX ON VENEREAL DISEASES. Crown 8vo, 82 Illustrations, 4s.

WILLIAM MARTINDALE, F.C.S.
AND
W. WYNN WESTCOTT, M.B. LOND.

THE EXTRA PHARMACOPŒIA with the additions introduced into the British Pharmacopœia 1885 and 1890; and Medical References, and a Therapeutic Index of Diseases and Symptoms. Sixth Edition, limp roan, med. 24mo, 7s. 6d.
[*Now ready.*

ANGEL MONEY, M.D., F.R.C.P.
Assistant Physician to University College Hospital, and to the Hospital for Sick Children, Great Ormond Street.

THE STUDENT'S TEXTBOOK OF THE PRACTICE OF MEDICINE. Fcap. 8vo, 6s. 6d.

A. STANFORD MORTON, M.B., F.R.C.S. ENG.
Assistant Surgeon to the Moorfields' Ophthalmic Hospital, &c.

REFRACTION OF THE EYE: Its Diagnosis, and the Correction of its Errors, with Chapter on Keratoscopy. Fourth Edition. Small 8vo, 3s. 6d.

C. W. MANSELL MOULLIN, M.A., M.D. OXON., F.R.C.S. ENG.
Assistant Surgeon and Senior Demonstrator of Anatomy at the London Hospital.

SPRAINS; THEIR CONSEQUENCES AND TREATMENT. Crown 8vo, 5s.

WILLIAM MURRAY, M.D., F.R.C.P. LOND.
Consulting Physician to the Children's Hospital, Newcastle-on-Tyne, &c.

ILLUSTRATIONS OF THE INDUCTIVE METHOD IN MEDICINE. Crown 8vo, 3s. 6d.   [*Now ready.*

WILLIAM MURRELL, M.D., F.R.C.P.
Lecturer on Materia Medica and Therapeutics at Westminster Hospital.

I.

MASSOTHERAPEUTICS; OR MASSAGE AS A MODE OF TREATMENT. Fifth Edition, crown 8vo, 4s. [*Now ready.*

II.

WHAT TO DO IN CASES OF POISONING. Sixth Edition, royal 32mo, 3s. 6d. [*Just published.*

G. OLIVER, M.D., F.R.C.P.
I.

ON BEDSIDE URINE TESTING : a Clinical Guide to the Observation of Urine in the course of Work. Fourth Edition, fcap. 8vo, 3s. 6d.

II.

THE HARROGATE WATERS : Data Chemical and Therapeutical, with notes on the Climate of Harrogate. Crown 8vo, with Map of the Wells, 3s. 6d.

K. W. OSTROM.
Instructor in Massage and Swedish Movements in the Hospital of the University of Pennsylvania.

MASSAGE AND THE ORIGINAL SWEDISH MOVEMENTS. Second Edition, With Illustrations, 12mo. [*Nearly ready.*

R. DOUGLAS POWELL, M.D., F.R.C.P., M.R.C.S.
Physician to the Hospital for Consumption and Diseases of the Chest at Brompton, Physician to the Middlesex Hospital.

DISEASES OF THE LUNGS AND PLEURÆ IN-CLUDING CONSUMPTION. Third Edition, with coloured plates and wood-engravings, 8vo, 16s.

FRANCIS H. RANKIN, M.D.
President of the Newport Medical Society.

HYGIENE OF CHILDHOOD : Suggestions for the care of children after the period of infancy to the completion of puberty. Crown 8vo 3s.

SAMUEL RIDEAL, D.SC. (LOND.), F.I.C., F.C.S., F.G.S.
Fellow of University College, London.

I.

PRACTICAL ORGANIC CHEMISTRY. The detection and properties of some of the more important organic compounds. 12mo, 2s. 6d.

II.

PRACTICAL CHEMISTRY FOR MEDICAL STUDENTS, Required at the First Examination of the Conjoint Examining Board in England. Fcap. 8vo, 2s.

E. A. RIDSDALE.
Associate of the Royal School of Mines.

COSMIC EVOLUTION: being Speculations on the Origin of our Environment. Fcap. 8vo, 3s.

SYDNEY RINGER, M.D., F.R.S.
Professor of the Principles and Practice of Medicine in University College, Physician to, and Professor of Clinical Medicine in, University College Hospital.

A HANDBOOK OF THERAPEUTICS. Twelfth Edition, revised, 8vo, 15s.

FREDERICK T. ROBERTS, M.D., B.SC., F.R.C.P.
Examiner in Medicine at the University of London; Professor of Therapeutics in University College; Physician to University College Hospital; Physician to the Brompton Consumption Hospital, &c.

I.

A HANDBOOK OF THE THEORY AND PRACTICE OF MEDICINE. Eighth Edition, with Illustrations, large 8vo, 21s.

II.

THE OFFICINAL MATERIA MEDICA. Second Edit., entirely rewritten in accordance with the latest British Pharmacopœia, with the Additions made in 1890, fcap. 8vo, 7s. 6d.

BERNARD ROTH, F.R.C.S.
Fellow of the Medical Society of London.

THE TREATMENT OF LATERAL CURVATURE OF THE SPINE. Demy 8vo, with Photographic and other Illustrations, 5s.

ROBSON ROOSE, M.D., F.R.C.P. EDIN.

I.

LEPROSY, AND ITS TREATMENT AS ILLUS-TRATED BY NORWEGIAN EXPERIENCE. Crown 8vo, 3s. 6d.

II.

GOUT, AND ITS RELATIONS TO DISEASES OF THE LIVER AND KIDNEYS. Sixth Edition, crown 8vo, 3s. 6d.

III.

NERVE PROSTRATION AND OTHER FUNC-TIONAL DISORDERS OF DAILY LIFE. Second Edition, demy 8vo, 18s. [*Now ready.*]

DR. B. S. SCHULTZE.

THE PATHOLOGY AND TREATMENT OF DIS-PLACEMENTS OF THE UTERUS. Translated by J. J. MACAN, M.A., M.R.C.S., and edited by A. V. MACAN, B.A., M.B., Master of the Rotunda Lying-in Hospital, Dublin. With Illustrations, medium 8vo, 12s. 6d.

WM. JAPP SINCLAIR, M.A., M.D.

Honorary Physician to the Manchester Southern Hospital for Women and Children, and Manchester Maternity Hospital.

ON GONORRHŒAL INFECTION IN WOMEN. Post 8vo, 4s.

ALEXANDER J. C. SKENE, M.D.

Professor of Gynæcology in the Long Island College Hospital, Brooklyn.

TREATISE ON THE DISEASES OF WOMEN. With 251 engravings and 9 chromo-lithographs, medium 8vo, 28s.

ALDER SMITH, M.B. LOND., F.R.C.S.

Resident Medical Officer, Christ's Hospital, London.

RINGWORM: ITS DIAGNOSIS AND TREATMENT. Third Edition, rewritten and enlarged, with Illustrations, fcap. 8vo, 5s. 6d.

JOHN KENT SPENDER, M.D. LOND.

Physician to the Royal Mineral Water Hospital, Bath.

THE EARLY SYMPTOMS AND THE EARLY TREATMENT OF OSTEO-ARTHRITIS, commonly called Rheumatoid Arthritis. With special reference to the Bath Thermal Waters. Small 8vo, 2s. 6d.

LOUIS STARR, M.D.
Clinical Professor of Diseases of Children in the Hospital of the University of Pennsylvania.

HYGIENE OF THE NURSERY. Including the General Regimen and Feeding of Infants and Children, and the Domestic Management of the Ordinary Emergencies of Early Life. Second edition, with illustrations, crown 8vo, 3s. 6d.

LEWIS A. STIMSON, B.A., M.D.
Professor of Clinical Surgery in the Medical Faculty of the University of the City of New York, etc.

A MANUAL OF OPERATIVE SURGERY. With three hundred and forty-two Illustrations. Second Edition, post 8vo, 10s. 6d.

JUKES DE STYRAP, M.K.Q.C.P.
Physician-Extraordinary, late Physician in Ordinary to the Salop Infirmary; Consulting Physician to the South Salop and Montgomeryshire Infirmaries, etc.

I.

THE YOUNG PRACTITIONER: With practical hints and instructive suggestions, as subsidiary aids, for his guidance on entering into private practice. Demy 8vo, 7s. 6d. *nett.*

II.

A CODE OF MEDICAL ETHICS: With general and special rules for the guidance of the faculty and the public in the complex relations of professional life. Third edition, demy 8vo, 3s. *nett.*

III.

MEDICO-CHIRURGICAL TARIFFS. Fourth edition, revised and enlarged, fcap. 4to, 2s. *nett.*

IV.

THE YOUNG PRACTITIONER: HIS CODE AND TARIFF. Being the above three works in one volume. Demy 8vo, 10s. 6d. *nett.*

C. W. SUCKLING, M.D. LOND., M.R.C.P.
Professor of Materia Medica and Therapeutics at the Queen's College, Physician to the Queen's Hospital, Birmingham, etc.

I.

ON THE DIAGNOSIS OF DISEASES OF THE BRAIN, SPINAL CORD, AND NERVES. With Illustrations, crown 8vo, 8s. 6d.

II.

ON THE TREATMENT OF DISEASES OF THE NERVOUS SYSTEM. Crown 8vo, 7s. 6d.

JOHN BLAND SUTTON, F.R.C.S.
Lecturer on Comparative Anatomy, and Assistant Surgeon to the Middlesex Hospital.

LIGAMENTS: THEIR NATURE AND MORPHO-LOGY. Wood engravings, post 8vo, 4s. 6d.

HENRY R. SWANZY, A.M., M.B., F.R.C.S.I.
Examiner in Ophthalmic Surgery n the Royal University of Ireland; Surgeon to the National Eye and Ear Infirmary, Dublin, etc.

A HANDBOOK OF DISEASES OF THE EYE AND THEIR TREATMENT. Third Edition, Illustrated with Wood Engravings, Colour Tests, etc., large post 8vo, 10s. 0d.

EUGENE S. TALBOT, M.D., D.D.S.
Professor of Dental Surgery in the Women's Medical College.

IRREGULARITIES OF THE TEETH AND THEIR TREATMENT. With 152 Illustrations, royal 8vo, 10s. 6d.

E. G. WHITTLE, M.D. LOND., F.R.C.S. ENG.
Senior Surgeon to the Royal Alexandra Hospital, for Sick Children, Brighton.

CONGESTIVE NEURASTHENIA, OR INSOMNIA AND NERVE DEPRESSION. Crown 8vo, 3s. 6d.

JOHN WILLIAMS, M.D., F.R.C.P.
Professor of Midwifery in University College, London; Obstetric Physician to University College Hospital

CANCER OF THE UTERUS: BEING THE HARVEIAN LECTURES FOR 1886. Illustrated with Lithographic Plates, royal 8vo, 10s. 6d.

BERTRAM C. A. WINDLE, M.A., M.D. DUBL.
Professor of Anatomy in the Queen's College, Birmingham; Examiner in Anatomy in the Universities of Cambridge and Durham.

A HANDBOOK OF SURFACE ANATOMY AND LANDMARKS. Illustrations, post 8vo, 3s. 6d.

DAVID YOUNG, M.C., M.B., M.D.
Fellow of, and late Examiner in Midwifery to, the University of Bombay, etc.

ROME IN WINTER AND THE TUSCAN HILLS IN SUMMER. A Contribution to the Climate of Italy. Small 8vo, 6s.

www.ingramcontent.com/pod-product-compliance
Lightning Source LLC
Chambersburg PA
CBHW021344210326
41599CB00011B/745